COLOR MANAGEMENT

imaging.org

COLOR MANAGEMENT

UNDERSTANDING AND USING ICC PROFILES

Edited by

Phil Green

London College of Communication, UK

A John Wiley and Sons, Ltd, Publication

Registered office
John Wiley & Sons Ltd, The Atrium, Southern Gate, Chichester, West Sussex, PO19 8SQ, United Kingdom

For details of our global editorial offices, for customer services and for information about how to apply for
permission to reuse the copyright material in this book please see our website at www.wiley.com.

Library of Congress Cataloguing-in-Publication Data

Green, Phil, 1953-
  Color management : understanding and using ICC profiles / edited by Phil Green.
       p. cm.
  Includes bibliographical references and index.
  ISBN 978-0-470-05825-1 (cloth)
 1. Image processing–Digital techniques–Standards.   2. File organization (Computer science)
3. Color photography–Digital techniques.   4. Color–Standards.   5. Ink-jet printing.   6. Colorimetry.
7. Photography, Orthochromatic.   I. Title.

  TA1638.G74 2010
  621.36'7–dc22

                                                                                          2009045131

A catalogue record for this book is available from the British Library.

ISBN: 978-0-470-05825-1

Set in 10/12 Times by Thomson Digital, Noida, India.
Printed in Great Britain by CPI Antony Rowe, Chippenham, Wiltshire.

Contents

About the Editor

Phil Green is a Reader in Color Imaging at London College of Communication, a constituent college of the University of the Arts, London, where he has worked since 1986. He specializes in color imaging, and runs a postgraduate program including both MSc and doctoral courses. Prior to commencing at LCC Phil worked for 14 years in the printing industry in London. He received a PhD in color science from the Color & Imaging Institute of the University of Derby, UK in 2003 and an MSc in Interactive Systems Analysis from the University of Surrey in 1995. He has published widely on related topics. He has authored and edited a number of textbooks on color, imaging and graphic arts, including Color Engineering, Understanding Digital Color, Digital Photography and Professional Print Buying, together with numerous papers in peer-reviewed journals and conferences. Phil is a member of the Society for Imaging Science and Technology and the Institute of Physics Printing and Graphic Science Group. He serves on Technical Committees of ISO and CIE, and became Technical Secretary of the ICC in 2005.

Series Editor's Preface

What is meant by managing color? The best way to get a practical feel about color management in imaging systems is to consider a few historical examples of both "closed" and "open" systems. The best example of a closed system would be silver halide color transparencies. In color transparency systems the manufacturer controlled all aspects of the color reproduction by specifying the spectral sensitivities of the silver halide emulsions (how the film sees colors), the dyes used to form the final colors, and the chemical processes that convert the silver halide to a black-and-white three color separation image (red, green and blue) and finally forms the color image (cyan, magenta and yellow dyes). The manufacturers specified the chemical process, but could not always "enforce" how the process was carried out; however for the most part the results were reliable. While the photographic color transparency imaging systems were closed, the color reproduction varied from film to film. Indeed there was a lot of "argument" about Kodachrome versus Ektachrome skies, Agfachrome reds and Fujichrome greens. All these films were based on the same principals, but they had different spectral sensitivities, used different dye sets, and in the early days used different chemical processes (which, in the long run, became a single chemical process based on Ektachrome). Professional and amateur photographers could pick the transparency film he or she liked best based on the color, sharpness and graininess, but there was little they could do to change the results.

The photographic color negative system presented the first bridge to a partially open color imaging system. Each manufacturer developed the color negative film, the color processing, and the color paper and processing in such a way that a consistent color reproduction could be obtained if one followed the directions carefully. Each color negative film differed in spectral sensitivities, image dyes, colored couplers (to help mask the unwanted absorptions of the image dyes) and various chemical interactions within the film to give better sharpness, less grain and more vivid colors. Again the professional or amateur could pick the film they liked best. The "open" systems aspect came from the darkroom where an advanced amateur or professional could use any number of techniques to alter the color and sharpness to meet their needs. These darkroom techniques were replaced in the 1990s by using digital scanners to translate the dye images of color negatives (and slides) into red-green-blue digital images, which in turn could be processed by the means of advanced image processing algorithms into "better" images for display or printing. Well before the desktop or photo-lab scanners used digital means to alter images, the graphics arts industry was using both analog scanners and digital scanners (very large, expensive devices) to make color separations from negatives or transparencies, which in turn were used to make printing plates. The selection of halftone patterns, printing technology, inks and papers for graphic arts reproductions led to a wide range of creativity and quality; more of an art than a science.

The advent and dominance of digital photographic imaging, color scanners, color copiers and color printers (based on ink, toner, thermal dyes, etc.) has led to the era of "open" color imaging systems. In the earlier closed or mostly closed color imaging systems, there were the obvious color failures noted by unhappy customers who complained that their faces were too red, the tablecloth is not purple but blue, the morning glory is a blue-purple and not pink, and other such concerns. Each of these failures could be accounted to some specific aspect of the closed imaging system and could only be corrected by new film design or a lot of darkroom manipulations. Today, with the proliferation of many digital imaging devices such as cameras, scanners and copiers (with their different color spaces like sRGB, RGB 64, Adobe RGB, etc.), and many imaging display devices including a diminishing number of CRT monitors, the dominant LCD monitor, many different TV displays and projectors, and a vast array of color printers (electro-photographic, ink jet, dye thermal transfer, etc.), each with their own colorants (and often specific papers for good results), the ability to control color quality has become a challenge if not a nightmare. Using a given LCD monitor, the same "scene" taken with three different cameras (of the same resolution) will have different color reproduction. Then using three different ink jet printers to print the three camera images will result in nine prints, none of which will have the same color preproduction. How does one solve this problem?

In the early 1990s a group of color scientists and engineers recognized the need for a formal approach to transferring color information between independent color devices. The subsequent version of the ICC Profiles, while an impressive start, failed to gather the required support from users and manufacturers alike. v4 of the ICC Profile cleared up the problems found in the earlier version and was adopted by ISO 15076. Today the challenges for ICC are twofold: (1) to consolidate the adoption of v4 and ensure widespread understanding of how to generate and use profiles; (2) to enhance the color management architecture and profile format in order to address needs not fully addressed in v4.

The ICC Profile goes a long way in solving this problem and this is the subject of the 9th offering of the Wiley-IS&T Series in Imaging Science and Technology: ***Color Management: Understanding and Using ICC Profiles*** Edited by Phil Green.

To understand the basic benefits of an ICC Profile, consider the following simple case. An image is recorded with a digital still camera that is calibrated to record color images using the sRGB color space. This means that the digital code in sRGB space can be directly related to some XYZ or Lab color defined by the CIE color matching functions, which act as a stable reference space for all colors seen by the "standard observer". Now this color image is to be viewed on a LCD display, which has been calibrated to the CIE system, say in Lab space. Hence the sRGB values of the digital camera can be converted to their respective Lab values and these Lab values can be converted to the digital values that drive the LCD display. Using the Lab color space as the common reference to both calibrations (camera and monitor), we can view the image as it was seen by the camera (with limits imposed by the sRGB color space and the limits of the LCD primary colors). Now say we wish to print the image. This can be done by using either the sRGB values or LCD digital values and converting back to the Lab values. These Lab values can be matched to the calibration of the printer that takes into consideration the type of halftone used, the colorants used to form the hardcopy and even the viewing illuminant. The calibration defines each RGB (or CYMK) value of the printer driver (the hardware and firmware in the printer) to the final Lab color value. Hence the Lab values of the image can be used (via interpolation algorithms) to generate the RGB or CYMK values of the printer to form the final image with the "same" colors seen by the camera. However, the user might like to change the color "intent" by moving from natural color (the original Lab values) to what is often called "vivid" colors where the color saturation is increased. Or it might turn out that some of the original sRGB values (transformed to Lab values) are beyond the gamut of the printer, so the print driver uses a gamut matching function to make the image look as natural as possible. The ICC Profile makes all this and much more possible. In short, the ICC Profile provides a systematic way to carry color information between a variety of "open" system color imaging devices.

Color Management: Understanding and Using ICC Profiles provides a concise and systematic description of ICC Profiles, the underlying color and color vision theory and how ICC Profiles are constructed. The process of creating an ICC Profile is complex and can be very confusing, but Dr. Green has taken out the confusion and provided an easy to follow process to generate the ICC Profiles. This text is an absolute must for color scientists and engineers who are involved in display and hardcopy technology. In addition, this text will be invaluable for all students and instructors who are learning or teaching the practical application of color reproduction in the digital age. The key issues covered in this text under the umbrella of the ICC White Papers are: (1) Understanding of different image states in a color reproduction workflow and the rendering intents appropriate to these states; (2) The use a of a reference gamut to remove ambiguities in interpreting source data when using the Perceptual rendering intent; (3) Correct interpretation of colorimetry in the Profile Connection Space, including the use of chromatic adaptation and requirements for display measurement; (4) Techniques for encoding and converting high dynamic range, scene-referred images using the profile format. ICC promotes wider understanding of these topics through the ICC White Papers, and through the ICC Developer Conference. This book provides a summary of current thinking in the ICC, written by the leading color scientists who make up the ICC membership.

Michael A. Kriss
Formerly of Eastman
Kodak Research Laboratories
and the University of Rochester

Preface

With the publication of a new version of the International Color Consortium (ICC) specification, it is timely to publish this collection of material based on the ICC White Papers. These documents contain high-quality material which has undergone extensive peer review within the ICC and between them provide a consistent and technically sound set of information and recommendations on color management.

Since 2005 I have had the tremendous privilege of acting as Technical Secretary for the ICC, and one aspect of this role is to answer questions on color management topics from visitors to the ICC web site (www.color.org/whitepapers.html). In doing so I realized that, while there is a great deal of excellent literature on color management, there was a need for an integrated text that organizes the White Paper material and is in sync with the latest version of the specification.

I should acknowledge here the lead authors of the original White Papers which are adapted in this book: Max Derhak (Onyx Graphics), Bob Hallam (Worldcolor), Jack Holm (Consultant), Tony Johnson (London College of Communication), William Li (Kodak), Ann McCarthy (Lexmark International), Craig Revie (Fujifilm and FFEI UK), and Ingeborg Tastl (HP). Draft White Papers and other documents were also contributed by the same authors and by Marti Maria (HP), David McDowell (Kodak and NPES), George Pawle (Kodak), and Robert Poe (Toshiba America Business Solutions).

The contribution of the many other ICC members who helped in developing both published and draft White Papers, and who provided comments on the edited versions presented here, should also be recognized. I apologize for being unable to thank all of them here, but I should in particular mention Harold Boll (Toshiba America Business Solutions), Nicolas Bonnier (Océ), Hitoshi Urabe (Fujifilm), Uwe Krabbenhoeft (Heidelberg), Marc Mahy (Agfa), Yue Qiao (Ricoh Americas Corporation), Steve Smiley (Vertis Communications), James Vogh (X-Rite), Eric Walowit (Color Savvy), and the current ICC Chair, Thomas Lianza (X-Rite), and the ICC Secretary, Kip Smythe (NPES).

Notes for the chapter on color stability in inkjet were prepared by Neville Bower (Felix Schoeller) and Phil Bowles, and Gregory High redrew many of the figures.

I am also grateful to my employer, London College of Communication, for allowing me to undertake work for the ICC while at the same time running the MSc Digital Colour Imaging course and supervising research students. My postgraduate students have always provided a spur to curiosity, invaluable feedback, and a test bed for new ideas.

The book has been a long time in gestation, and I must thank Project Editor Nicky Skinner and Commissioning Editor Georgia Pinteau at John Wiley & Sons, Ltd for their unending patience and assistance.

Finally, I would like to express my appreciation to my partner, Ruth, and daughter, Rosalie, who make the world more colorful.

Part One

General

1

Introduction

ICC White Papers are one of the formal deliverables of the International Color Consortium, the other being the ICC specification itself – ISO 15076: *Image technology color management – Architecture, profile format, and data structure.* The White Papers undergo an exhaustive internal development process, followed by a formal technical review by the membership and a ballot for approval by the ICC Steering Committee.

The White Papers generally address single topics within color management and the use of the ICC profile, but together they include the collected wisdom and consensus view of a community of leading color scientists and developers who represent all the major companies active in the field of color management. The White Papers are based on well-founded color science, concrete experience, and best practice.

Color Management is mainly based on the ICC White Papers, including those already published on the ICC web site and draft versions published internally. The chapters here represent edited, updated, and sometimes expanded versions of the documents that have been published by the ICC. In many cases the White Paper on which a chapter is based is still in development, and this book represents an opportunity to provide an insight into the material which is undergoing discussion. Unlike the published White Papers, the chapters in *Color Management* have not been formally approved by the ICC, and it must be emphasized that I am entirely responsible for any errors, ambiguities, or misinterpretations.

Color Management also includes the chapter "ICC Profile Mechanics," which is not based on a White Paper but on material presented at the ICC Developer Conference in Portland, Oregon, in November 2008, by Marti Maria of HP.

The recent approval and publication of the revised ICC Version 4.3 specification (also published as ISO 15076-1:2010) is an important step in the evolution of color management, since it represents a significant improvement in the clarity and consistency of the specification and incorporates all the amendments approved by the ICC between publication of the first Version 4 specification in 2001 and June 2009. By marking this new version of the specification by the present volume, I hope that the path to adoption and implementation can be eased by the expert guidance contained in these chapters.

Color Management: Understanding and Using ICC Profiles   Edited by Phil Green
© 2010 John Wiley & Sons, Ltd

The ICC White Papers are written primarily for the main segments of the color management community: the scientists and developers who devise and maintain color management hardware and software, the professional users who implement color management solutions, and those general users who would like to understand more about how to get colors to come out right. While the underlying content remains the same for all these segments, the technical level, language, and style in the different White Papers vary considerably, and I have intentionally preserved the approach taken in the original paper in each case. Readers will therefore find no particular consistency of voice or technical level between the White Papers or the chapters here.

While many of the chapters incorporate a definition of key terms used, readers should also consult the Glossary in Chapter 8 for explanations of any term. As a result of the ICC internal review process, the use of terminology should be reasonably consistent throughout the White Papers and this book.

The content of the book is organized into five main parts. In the first part general material about the ICC architecture, the profile format and its history, and the future of the ICC architecture are discussed. The second part focuses on issues around Version 4 of the specification, and in particular on the v4 perceptual intent. In the next part a range of workflow issues are discussed, including those specific to digital photography and graphic arts. The fourth part addresses a range of topics around measurement and viewing conditions for color management, while the fifth part gives detailed guidance for profile creators. As with the original ICC White Papers, the level of the chapters ranges from introductory to advanced. Together they provide both conceptual information and practical recommendations to users and implementers of color management systems.

The ICC specification and the White Papers will continue to be developed, and readers are recommended to visit the ICC web site at http://www.color.org for the latest versions of these documents, together with numerous other resources posted on the site for the benefit of the color management community.

The ICC is a member consortium, open to any organization willing to pay the membership fee and abide by the member agreement. ICC membership confers substantial benefits, and organizations which have an interest in color and are not already members should consider joining and contributing to the development of color management in the international arena.

Some readers may be puzzled by the apparent inconsistency between the European conventions adopted in spelling and notation in the latest version of the specification and the US spelling used in many ICC documents and in type names in the specification. Less obvious is that there has been a considerable input from ICC members in Japan, who in translating the specification into Japanese have identified many ambiguities and errors in the use of English in the previous version. Since the specification has become an ISO standard, the latest version has consistently adopted ISO conventions for spelling and notation, while retaining the original type names. ICC itself is an international organization and has not attempted to agree spelling conventions (there are invariably more interesting things to discuss!), and as a result the White Papers and other documents on the web site use both US and UK spellings, usually depending on the provenance of the original authors. In *Color Management* I have converted spellings to the US form, with the exception of type names from the specification.

Considerable effort has gone into preparing the guidance provided in the ICC White Papers and the versions in *Color Management*. Nevertheless, readers should ensure that the use of the material meets their needs, and neither I, the ICC, nor the publisher accept any liability for losses suffered as a consequence of any of the information or recommendations given.

2

Color Management – A Conceptual Overview

This chapter provides an overview of some of the key concepts in color management, and describes the evolution from v2 of the ICC specification to the v4 specification in use today. It summarizes the color rendering and re-rendering options provided by the different rendering intents within an ICC profile.

Color management can be defined as the "communication of the associated data required for unambiguous interpretation of color content data, and application of color data conversions as required to produce the intended reproductions."

Color content may consist of text, line art, graphics, and pictorial images, in raster or vector form, all of which may be color managed. To be successful, color management must consider the characteristics of input and output devices in determining the appropriate color data conversions for these devices.

2.1 Evolution

We can identify four distinct phases in the evolution of the understanding of color management. Initially there was what could be described as "digital color mode," whereby color was expressed in terms of the coordinates obtained on devices, in color spaces such as RGB, CMYK, and YCC. Subsequently it became common to describe colors by means of their colorimetry, using the well-established CIE system. The move to colorimetric specification of color led to the notion of "device-independent color" – the idea that a color could be expressed in terms of its colorimetry independently of the device used to create it. Communicating color through CIE colorimetry works well when the viewing conditions are well defined and the output device is fixed, so that there is a well-defined process for rendering color to the output system and its viewing condition (as is the case, for example, with the television model).

In more recent years, the effect of the viewing conditions on the appearance of color has become more widely appreciated, together with the effect that this has on the desired colorimetry of a reproduction. However, the human visual system is still not fully understood, and although we have models such as CIECAM02 which are successful for certain viewing conditions, there are as yet no published models that provide a robust and comprehensive description of appearance.

Today our understanding is based on the different states in which an image can exist. The desired appearance of an image depends on the output medium and its viewing conditions, and some form of rendering is required to transform an image from one image state (such as scene-referred colorimetry) to another (such as output-referred display or print). This concept of rendering is distinguished from gamut mapping, which can be thought of as primarily an operation to clip a source gamut to a destination gamut of a different (usually smaller) size. The media- or image-specific preference aspect of the mapping can therefore be considered more as an operation to render between different image states.

If we are to successfully render between image states, it is essential that we are able to unambiguously interpret color data and hence it is necessary that the image state at any point in the workflow is known. This type of approach is in fact implicit in traditional photography and graphic arts, where for example a transparency is interpreted in a certain way in order to obtain a pleasing reproduction on a print.

We can then define two types of color management workflow. In the first, we can consider the output device to be fixed, and thus the intended viewing conditions and mode of viewing, the dynamic range and gamut of the reproduction, and other characteristics of the medium such as the substrate and the type of surface, are all known. In this case we can ship the desired colorimetry to the output device, usually by means of a colorimetric transform to the device encoding.

In the second type, the output device is not completely fixed but is variable in some way (e.g., through the option of having different viewing conditions, or through different output media being available). In this case the optimal image appearance may be device dependent, and a successful cross-media or cross-device color transform includes a color rendering between different image states.

2.2 Color Appearance

Appearance models are frequently useful in imaging applications. Transforms between corresponding colors in different viewing conditions often apply the chromatic adaptation component of a color appearance model. Appearance models also provide more perceptually uniform spaces for gamut mapping, and can be used to model the dependence of colorfulness on absolute luminance. Some device characterization methods also perform error minimization in color appearance coordinates.

However, since the cross-media objective is often *not* to reproduce appearance, color rendering approaches that independently use appearance models to deal with viewing condition differences, and gamut mapping to deal with gamut differences, may not be optimal. The primary color rendering task may actually be to alter appearance in order to produce a pleasing reproduction on different media. The changes in colorimetry driven by the appearance model may then be counter to those driven by gamut mapping, making independent optimization ineffective. Moreover, we do not yet have models that robustly describe color appearance, particularly for complex images as opposed to uniform stimuli.

2.3 Reproduction Models

Reproduction models have to consider simultaneously the effects of viewing condition, media limitations, user preferences, and, potentially, image characteristics in developing optimal color rendering transforms. Such models can be based on an analysis of what is done to image colorimetry by experts in achieving excellent cross-media reproductions. They are thus at least partially empirical – but so are appearance models and gamut mapping algorithms. They can add components based on our understanding of the human visual system as this understanding develops. The key to a successful model is simultaneous optimization of all the parameters described above.

2.4 Color Imaging Architecture

Unambiguous exchange of color image data requires that the different attributes of color are well defined. ISO 22028-1 provides definitions of color space encoding, viewing conditions, image state, and reference medium.

Color rendering can be applied in either proprietary or standardized ways. Standardization, where applicable, is essential in reducing possible ambiguity, but it should also be recognized that proprietary methods have the potential for adding value and providing enhanced implementations.

Implementation mechanisms should be aimed at producing standard color encodings (i.e., encodings of the colorimetry of an image on a reference medium, including the associated viewing conditions). An image writer or reader is then required to color-render to or from this standard color encoding. Attaching a color profile provides the transforms to be applied to the encoded image data in order to produce image colorimetry in a profile connection space (PCS) describing a specified medium (including its associated viewing conditions). Appropriate transforms to and from the PCS are linked by the color management module (CMM).

2.5 Color Rendering Options

In a color reproduction workflow, there are two possible options for handling the color rendering. An intermediate reproduction description provides input-side color re-rendering to some well-defined real or virtual reference medium. Image data is then exchanged and output-side color re-rendering is performed from the reference medium to the actual output medium. Alternatively, a deferred color rendering is achieved by encoding source colorimetry with the medium characteristics and information about the viewing conditions. The color re-rendering capability must be made available at the output stage, so that when final output is selected, color re-rendering is performed directly from source to actual output.

Early binding and late binding are terms used in graphic arts to designate when in the workflow the conversion/separation to the printing process colors cyan, magenta, yellow, and black (CMYK) takes place. This workflow usually starts with an intermediate reproduction description created on a computer or produced by a capture device (now almost invariably red, green, and blue, or RGB, although capture directly to CMYK is possible).

Early binding produces an intermediate reproduction description, based on some assumed output device. This (second) intermediate image may need to be color re-rendered to different output devices and media, such as proofs and prints made by different printing processes. It is helpful if early binding images are in some "standard" CMYK color encoding.

Late binding defers the conversion (or "separation") to device values until the actual output device is known. In this case, multiple files may be produced for the different devices.

The advantages of the intermediate reproduction description can be summarized as:

- Output is more consistent than with scene-referred exchange (since the desired artistic intent can be communicated in the intermediate image).
- Proven in practice by photographers and graphic artists.
- Commonly used bridging transforms for color re-rendering can be highly tuned and made widely available.
- Requires less sophisticated processing capability at output.

The disadvantages of the intermediate reproduction description are:

- Color re-rendering to actual output may be necessary.
- May not produce optimal results, particularly if the intermediate image reference medium is very different from the actual output medium.
- There is less output-side control of scene-to-picture color rendering.
- In the early binding case, assumptions that device values and gray component replacement (GCR) will or should be maintained when re-purposing may not be correct.

Deferred color rendering has the following advantages:

- Output-side control of color rendering and re-rendering are increased.
- Color rendering or re-rendering is direct to the actual output.
- There are no concerns that the intermediate image is too different from the actual output.
- In the late binding case, decisions involving device value selection (spot color substitution and solids) are deferred until the actual device is known.

The disadvantages of deferred color rendering are:

- Less consistent output due to greater color rendering freedom.
- A mechanism is needed for preview or proof of the color rendering.
- The image creator's artistic intent may not be maintained.
- Image data after processing for output is device specific, and can cause difficulties if fed back into open workflows.
- The capability to perform color rendering or re-rendering from the source encoding must be available at output.
- More hand tuning may be required, if more aggressive automated color rendering and re-rendering algorithms do not produce the desired result.

2.6 The Current Situation

In the case of color rendering (i.e., direct from scene to output-referred image data), the intermediate reproduction description approach dominates today. In most cases this is a standard output-referred exchange, using color encodings such as sRGB, Adobe RGB (1998), and ROMM RGB. Manually guided deferred color rendering (e.g., camera raw) is becoming increasingly popular, especially in professional markets, although even in this case color rendering is normally to a standard output-referred color image encoding for exchange. Here the concept of the digital negative and positive, in which a master file is archived for subsequent rendering, is relevant.

In the case of color re-rendering (i.e., from an image in one output-referred medium to a reproduction on another output-referred medium) both early and late binding workflows are used, although the image state is not always communicated. Re-rendering may be performed either by the CMM (using the media-relative colorimetric intent with black point compensation to scale the dynamic range of the first image state to that of the second) or by the profile (using the perceptual rendering intent to adjust both dynamic range and colorfulness to provide a preferred reproduction for the second medium).

ICC v2 profiles are limited in the performance and reliability of color re-rendering using the perceptual intent, primarily because the dynamic range and color gamut of the first image state is undefined when applying the profile to perform the re-rendering. This problem is addressed in ICC v4, which has a specified black and white point for the perceptual PCS and a well-defined Perceptual Reference Medium Gamut, although v4 profiles are not yet used in all workflows.

Using the media-relative colorimetric intent with black point compensation with v2 profiles deals with at least the first-order dependency of the desired appearance on the intended reproduction medium by means of the dynamic range adjustment, but this approach is not entirely optimal. Advanced CMM-based color re-rendering can overcome this limitation, but the use of such CMMs is not yet common and their required behavior is not standardized. The algorithms required to perform color re-rendering are rapidly evolving, and in some cases, with particularly difficult mappings between color gamuts, the transform must be hand tuned in order to achieve optimal performance.

For both color rendering and re-rendering, there are two types of implementation in use:

- sRGB is based on an output-referred intermediate reproduction description based on a reference display and viewing conditions.
- ICC profiles offer several rendering intents, supporting both color rendering and re-rendering.

sRGB is widely used, especially in consumer devices, and the quality of implementations continues to increase as understanding evolves and the color rendering capability increases.

ICC profiles provide a perceptual intent based on a reference print intermediate, together with measurement-based colorimetric intents which enable deferred color rendering by smart (generally proprietary) CMMs. They also enable colorimetric proofing. A degree of standardized color rendering capability is provided by some CMMs through support for media-relative colorimetric with black point compensation.

The ICC saturation intent enables proprietary workflows, where the rendering goal is different from that expressed in the perceptual and colorimetric intents.

Like sRGB, ICC-based color management is evolving as the understanding of the use cases, requirements, rendering methods, and color management architecture continues to increase.

2.7 ICC v2 Issues

Version 2 of the ICC specification had a number of significant shortcomings:

- Although the PCS is D50, the chromatic adaptation which had been performed to obtain a media white point in D50 was not required to be defined within the profile, and as a result the chromatic adaptation state of input data was ambiguous.
- The color re-rendering that was required in order to obtain the desired appearance on the PCS reference medium from input data was not defined, and for the perceptual intent there was no standard reference medium. This led to different assumptions about the PCS perceptual dynamic range and color gamut by different profiles.
- Colorimetric intents were not required to be measurement based, and since in addition measurement methods were not always well defined, the behavior of the colorimetric intents was unpredictable.
- There was insufficient flexibility in the transforms and color processing models provided within the v2 specification.

As a result of these shortcomings, capability limitations and interoperability problems could result.

There were at least three possibilities for input-side color re-rendering in v2:

1. Colorimetric with no black scaling
2. Colorimetric with black scaling
3. Perceptual to some arbitrary reference medium.

Depending on the source image, and the input profile re-rendering, the PCS colorimetry could thus be appropriate for a variety of different media and viewing conditions which were actually used within the profile.

The different input-side color re-rendering possibilities are illustrated in Figure 2.1.

These multiple input-side re-rendering possibilities lead to a dilemma for v2 output profiles. The perceptual intent of a v2 output profile was supposed to perform a pleasing re-rendering of the PCS image colorimetry to the actual output medium and viewing conditions. However, the output profile creator had no knowledge of the medium and viewing conditions for which the PCS colorimetry was appropriate! It is impossible to create an optimal perceptual rendering without this knowledge, and therefore optimal cross-vendor interoperability is precluded – while the output profile knows the end result, there are in effect many possible starting points in the PCS for a given set of input data, as illustrated in Figure 2.2.

The colorimetric rendering intent in v2 also presents implementation issues. In a v2 profile, the source colorimetry may be black scaled or color re-rendered to a proprietary reference

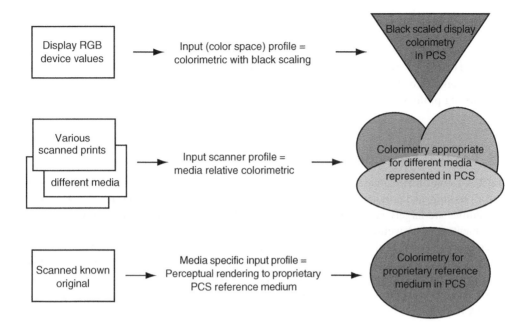

Figure 2.1   The v2 input color re-rendering possibilities

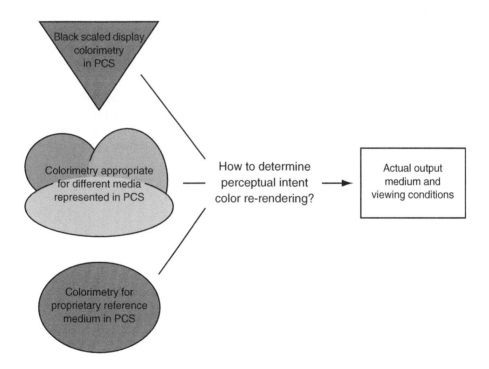

Figure 2.2   An output profile perceptual rendering intent has many possible data sources

medium, in order to enable improved interoperability within a single vendor's products. Because PCS colorimetry may not be accurate relative to the original, the CMM cannot rely on the source colorimetry as represented in the PCS, and as a result v2 profiles will not support advanced CMM color rendering. There are also other issues that arise with v2 profiles because of the ambiguity of the v2 specification and incorrect interpretation of the specification in constructing profiles.

2.8 The ICC v4 Solution

In ICC v4, colorimetric rendering intents are measurement based. They can therefore be relied on for proofing, and provide accurate colorimetry for CMM color re-rendering. Specification ambiguities are largely resolved and the text clarified to reduce the occurrence of incorrect implementations. A well-defined reference medium for the perceptual intent, with an associated gamut known as the Perceptual Reference Medium Gamut (PRMG), ensures cross-vendor interoperability. There is also greatly increased transform capability through extended look-up table (LUT) definitions, such as the lutAtoBtype which incorporates an additional matrix and curve and provides greater mathematical flexibility and an improved definition of 16-bit CIELAB.

2.9 ICC v4 Perceptual Intent

Significant improvements have been made to the interoperability of the v4 perceptual path. The v4 perceptual intent color reproduction path is illustrated in Figure 2.3. With the PRMG, both input and output profiles can be based on a well-defined intermediate image colorimetry appropriate for the PCS reference medium and viewing conditions. The task of the CMM is thus to connect profiles with the same (or very similar) PCS gamuts, and minimal gamut mapping is required because the image colorimetry in the PCS is matched for the input and the output. Differences between source and output media color gamut and viewing condition are then dealt with consistently within the mapping to or from the reference medium performed by each profile.

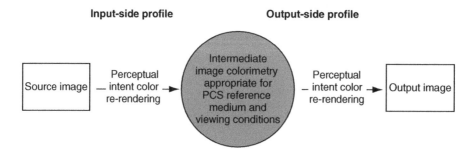

Figure 2.3   Perceptual intent color reproduction path in ICC v4

The v4 perceptual transform includes both the data (typically device value) to PCS colorimetry transform, and color re-rendering to and from the reference medium in the PCS. The re-rendering operation includes consideration of:

- differences in viewing conditions between source and reproduction and their appearance effects;
- differences in media characteristics and image state;
- color rendering preferences and the attributes of the preferred reproduction on the output medium.

If the profiles incorporate all of these considerations, the task of the CMM is simply to connect the profiles together to create the transform between source and output data.

The v4 perceptual transform is useful for general image reproduction across all devices and media. Since color re-rendering operations are typically proprietary, profiles from different sources may produce different "looks." Users can then select profiles based on color re-rendering preferences. This was difficult before v4 due to the issues with the v2 specification described above and a lack of coordination between the different color management components (the operating system, the application, and the driver and/or output system raster image processor (RIP)). As differences between actual and reference media decrease, the perceptual and colorimetric intents should converge. Before v4, users were cautious about the perceptual intent because of the inconsistencies with v2. However, it is still important that v4 profiles are correctly constructed and that color management is well coordinated in order to maximize the confidence of users.

2.10 ICC v4 Colorimetric Intents

The ICC v4 colorimetric path is illustrated in Figure 2.4.

The color gamut mapping performed by a v4 profile has three requirements:

1. The input data colorimetry should not be changed within the intersection of the input and output media gamuts.
2. Colors that are outside the source image gamut should not be produced in the output image.
3. Colors in the source image that are outside the output image gamut should be clipped.

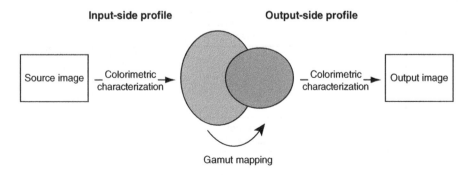

Figure 2.4   ICC v4 colorimetric path

A colorimetric transform includes the device data to PCS colorimetry transform, based on measurements made using standard methods (as defined in ISO 13655 and described in Chapter 20 below). The transform also includes chromatic adaptation to and from the D50 PCS white point, if the data has a different reference white. This allows gamut mapping to be performed directly, if desired. In proofing situations, the extent of gamut mapping required is best minimized by the choice of proofing media. As the chromatic adaptation matrix is included in the profile, it is invertible if CMM-based chromatic adaptation is desired. The colorimetric intent does not include other appearance transforms, in order to avoid unnecessary color appearance model complexity, instability, and other issues mentioned above.

Colorimetric transforms are useful for preview and proofing applications, and in support of CMM-based color rendering. The media-relative colorimetric with black point compensation (MRC + BPC) provides a standard baseline CMM color rendering that is adequate when the media, substrate, and gamut shape differences are not large. This baseline reproduction model includes chromatic adaptation and media white relative colorimetry with black point scaling (on XYZ coordinates). It also includes gamut expansion and compression as required. The current widespread use of MRC + BPC demonstrates the importance of media considerations.

2.11 ICC v4 CMM Color Rendering

In ICC v4, it is possible for color rendering to be performed by the CMM rather than the profile, as illustrated in Figure 2.5.

In this scenario, CMM algorithms color re-render source image colorimetry to be appropriate for the actual output medium, taking into consideration source and output medium color gamuts and viewing conditions. They can also support color appearance model-based color re-rendering. CMM-based color rendering can take advantage of full output medium gamut, and facilitate user adjustment of color re-rendering at the time of output. For more details on CMM capabilities in ICC v4, see Chapters 6 and 31 below.

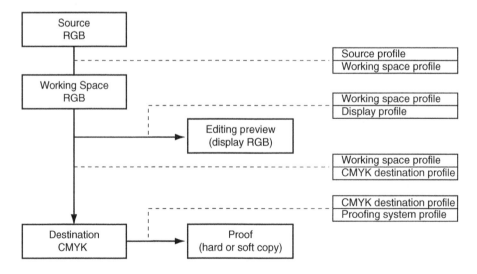

Figure 2.5   ICC v4 CMM color rendering

2.12 Moving Forward

Current research into color rendering supports both automated perceptual intent transform generation and selection, and increased CMM color rendering capability.

High-quality ICC v4 tools and profiles need to become more widespread to move completely away from v2 issues. However, it is acknowledged that considerable work is still needed to fully coordinate color management across operating systems, applications, and devices. User interfaces also need considerable work, but should be based on v4 solutions rather than on codifying v2 problems, and should ideally advance both color management and user interface effectiveness. ICC and its members and the color management community need to work in a coordinated way to advance all of the technologies described above, building where possible on solid understanding and communication. Clear and unambiguous definitions of color encoding and image state, for example through ISO 22028-1, are key elements of this process.

3

The Role of ICC Profiles in a Color Reproduction System

3.1 Introduction

Color reproduction can be a complex process. There are many different color reproduction industries, often utilizing different media from one to another, and within some industries there may well be multiple media used. These different industries will often have differing reproduction requirements, even for the same image, depending on the reproduction process itself and the stage in the workflow at which the reproduction is made. For example, an image on a computer display may be required to accurately match the color of the original image, or be a pleasing (idealized) reproduction of that image, or be a color match to a printed reproduction of the original (soft-proofing), which in turn may be a color accurate or a pleasing rendition of the original. One of the most important decisions that has to be made by a user is what kind of reproduction is required at each stage of a workflow where a digital image is rendered in some form.

In many systems where color reproduction is limited to a small number of input and output devices, the mathematical transformation which has to be applied to the image data to achieve the desired color is often heavily optimized for each device pair. In such situations the color reproduction requirements of each stage of the workflow are usually well defined, and the transformations are optimized for those requirements and the devices used. Although ICC profiles can be, and are, used to replicate such systems, many of them are based on proprietary algorithms – often utilizing measurement equipment specific to the system. In addition, the profiles may well provide procedures for fine tuning the algorithms. In the hands of reasonably skilled users these systems can – and do – produce results of very high quality.

However, in workflows where multiple devices might be used, and particularly where the devices may not be known at the time of image capture or generation, proprietary systems are often impractical. It was primarily for such workflows that the specification for ICC profiles was established. Its goal is to provide a mechanism for defining the color of image data in a way that makes it possible to exchange images between systems, while retaining any color requirements imposed on the image. However, it needs to be recognized that ICC profiles

Color Management: Understanding and Using ICC Profiles   Edited by Phil Green
© 2010 John Wiley & Sons, Ltd

do not, by themselves, comprise a color reproduction system. An application that provides color management is required to utilize them – and each application may provide different levels of functionality in order to meet the particular requirements of the users in the market sector that the product is serving. So, as long as the user selects an imaging application appropriate to their needs, it should be possible to use ICC profiles to provide the desired color reproduction.

Despite this, the ICC is sometimes criticized for various inadequacies of color reproduction. While some of the issues raised may be appropriate for attention by the ICC (and in most cases are being worked on by ICC Working Groups), others are more the responsibility of applications that use ICC profiles. In many cases, limitations and deficiencies encountered by users are those of the implementation, as opposed to the ICC specification. Some of the color reproduction issues are so dependent on the industry sector in which the images are being used that general solutions must be the responsibility of vendors and experts with experience in those markets. The ICC is a loose consortium of companies accommodating multiple industry sectors, and in many cases color reproduction solutions appropriate for one sector are not appropriate to others. Thus the ICC sees its main role as providing an open method to describe the color for each pixel of an image that needs to be matched, and a procedure for achieving that match. Where such a match is not desirable, the "best" solution is very difficult to define as it can depend on many factors. Thus, the best that the ICC can really offer are mechanisms (known as perceptual and saturation rendering intents) to enable a user to define that solution in a way that allows it to be communicated to others in the workflow, but not attempt to define how it is achieved. For this it is important that applications are selected that provide results suited to their needs.

The correct use of the optimized perceptual and saturation renderings within each industry sector enables the production of high-quality reproductions, tailored to the user's needs in that sector. These intents, together with the colorimetric renderings, enable many reproduction requirements to be met, and where an extension to the system is required for particular industry needs, vendors provide very sophisticated color reproduction systems. Such systems are based on the ICC specification, but include the additional tools demanded by the industry sectors they serve. Other vendors provide simpler systems that are easier to use, which serve other markets, often utilizing only the basic ICC architecture. Users need to verify that they are purchasing a system appropriate for their needs.

3.2 ICC Profiles – What Are They and How Are They Used?

Each ICC input (or source) profile provides a number of color transformations (in the form of look-up tables, matrices, and/or curves) that define the color expected from the encoded data of the digital image, in an open format. In other words, the profile defines the color to be expected with any set of image values – which are often device values, but may be in some standard color image encoding (such as sRGB). The color space used by ICC profiles is the internationally accepted CIE system for defining color matches, so by using this it is possible to ensure that colors from input will match those on output (assuming the output has an adequate color gamut), for the viewing conditions for which the color is defined. The conditions selected by the ICC are those defined in international standards for viewing transparencies and prints; the resultant color space is known as the profile connection space (PCS). The fact that the format is public means that any ICC-compliant system should be able to use these profiles to interpret the color intended for that digital image. In conjunction with the correct display and/or output (destination) profiles, various reproduction options can be achieved.

The reason why a profile contains multiple transforms is to allow the user to select the one appropriate for the purpose. The various rendering intents that these transforms provide are intended to be applicable to different reproduction goals. The choice can have a significant effect on the color reproduction achieved, so the selection of the appropriate transform is an important decision for the user.

The basic way in which ICC profiles are typically used to achieve color reproduction is by combining a source profile with a destination profile to enable input data to be transformed to that required to give the required color on output. Selection of the appropriate transforms, by selection of the rendering intent, enables the desired reproduction to be achieved. The combining of the profiles is performed by a CMM, which can be provided at various places in the workflow (such as the image editing software, raster image processor, or printer driver, among others). In some reproduction procedures there may be more than two profiles used (such as simulating a print on a display), or even special cases where only one is used that has been constructed by combining a source and destination profile (DeviceLink profiles). However, these are natural extensions of the basic procedure described here and greater detail will be found in the ICC workflow guidelines.

3.3 ICC Profiles as Part of a Color Reproduction System

Simply using a CMM that only supports the basic ICC architecture to calculate and apply the transformation from input device space to output device space does not necessarily provide a color reproduction system that suits all needs. So long as the application providing the CMM allows the selection of the appropriate rendering intents at the time when the appropriate profiles are combined, there are many market sectors where it is perfectly adequate – particularly where input devices are "smart." However, there are other markets where it may not be. In such situations additional functionality needs to be provided by the color management vendor.

3.3.1 Image Editing

One issue is that many captured images are not ideal. They frequently exhibit color casts, limited dynamic range, or poor tonal rendition, which may not be obvious on some media but will be when reproduced on others. Such "errors" need correcting during the process of reproduction. Algorithms for automatically optimizing digital images have been developed, and are a part of many image capture, color management, or editing applications. In fact they may often be applied without the user knowing. However, because of the subjective nature of color reproduction, such automatic algorithms may not suit every user, or every image. Thus, for high-quality imaging, unless the user is confident in the quality of captured images, every image should be assessed and corrected as necessary. Such corrections require a subjective assessment of the image, which means that it has to be rendered in some form to judge its quality. For many users a well-calibrated video display is adequate for this purpose, though for some high-quality applications the image is first rendered in its final form, which implies some sort of iterative correction process.

Each ICC profile is defined for a specific combination of device and media (as appropriate) and as such, when used appropriately, should enable faithful reproduction of the colorimetry of the encoded image. Although the perceptual and saturation rendering intents include optimizations for media and viewing condition differences, device profiles – which are determined independently of any images – do not apply image-specific optimizations. Where precision is of the utmost importance, color management software can be designed to update device profiles to also include image corrections, but because of the subjective nature of this correction it is usually sensible, in the view of many experts, to keep the characterization and image enhancement algorithms conceptually separate. Alternatively, the algorithms for image correction, if automated, can be applied at the same time as the media transform specified by the device profile. As "smart" CMMs (which add functionality by interpreting both profile and image information in calculating the reproduction transformation) are developed, such procedures are very likely.

An input profile can be embedded in an image, or sent as a separate file. Either way it can be used to define the intended color as already stated. However, the sender of the file has to be responsible for ensuring that the correct profile is embedded, but equally importantly has the responsibility for ensuring that the image is pleasing. If the image needs correction this should be undertaken prior to sending it, by directly editing either the image or the profile. In the event that this has not been done, and it is the responsibility of the receiver to optimize the image to make it pleasing, this must be made clear when the image is sent. The sender of the file must then be prepared to accept the changes made, or ensure that a proofing cycle that will enable corrections to be specified is part of the workflow.

3.3.2 Rendering Issues

The choice of rendering intent is an important one, as already discussed. General guidelines as to which rendering intent is appropriate to different types of images, and/or workflow stages, are given elsewhere in this book. Essentially the selection comes down to whether a colorimetric match is required between input and output, such as for proofing and preview applications (or when the output media have a gamut close to that of the image) or whether the reproduction is to be the most pleasing by compensating for the differences in viewing conditions and gamut between source and destination media.

The different rendering intent transforms in a profile are usually dependent on the profile creation software used to make them. While colorimetric renderings may well be somewhat different – because different vendors can use different targets for profile creation, different measurement devices, and different mathematical models – such differences are usually small. However, the perceptual and saturation intents can vary significantly. With older profiles there was an additional complication concerning ambiguity around the definition of the white and black values in the PCS to which the appropriate image data should be mapped, which could be interpreted differently by different profiling vendors. Thus, when profiles from different vendors were combined, the results could be unpredictable and/or low in quality. Although the use of Version 4 profiles should avoid this latter issue, it is not intended to ensure that the perceptual and saturation rendering intents provided by different vendors produce the same transformation. This is an area where different profiling vendors will provide solutions most appropriate to the markets they have most experience of, and it is up to the user to select that product which produces the most appropriate tables for their needs. The same vendor may even offer the option of different perceptual renderings to produce different "looks."

Differences between profiles will usually be more noticeable where the difference between the source and destination gamut is large. To enable consistency of rendering on the input side, the ICC suggests the Perceptual Reference Medium Gamut as a rendering target for the perceptual rendering intent. If this is used in a rendering workflow, the output profile does not have to make arbitrary choices about how it maps the source gamut to the output medium gamut.

One of the complications in trying to specify perceptual or colorimetric renderings in any objective way is the fact that there is limited agreement between experts as to what constitutes an optimum color re-rendering, which includes appearance and preference adjustments, and gamut mapping. This is complicated by the fact that such studies are inherently difficult. From the discussion above it will be clear that both media differences and image content affect the perceived quality of the color re-rendering, and separating these in any study is not easy. If both are included in the study it will generally be necessary to evaluate large numbers of images (maybe several hundred) before coming to a reasonable conclusion as to an optimum algorithm. The ICC sRGB v4 profile, for example, went through exhaustive testing by ICC members before it was adopted as a recommended solution to the perceptual intent transform from sRGB to the Perceptual Reference Medium Gamut.

Even if we assume that the image has been edited to remove any problems – so that the profiles are only expected to optimize the mapping for the media differences – it is still difficult to get agreement on that mapping. Trying to find a single algorithm that will work well for a variety of source and destination media types, and for a range of gamut shapes, complicates that further. All these reasons, together with the fact that other issues (e.g., viewing condition differences and user preferences) are often compensated for in perceptual and saturation renderings, make it very difficult to come to any general recommendation on the way to perform such mappings. In general, users with high-quality expectations must choose their color management software with care, or rely on expertly designed systems provided by companies for specific markets. Such systems may well provide correction routines to enable users to achieve specific rendering of particular colors.

3.3.3 Retention of Separation Information

One of the problems encountered in many practical color reproduction procedures is the difficulty of optimizing non-colorimetric profiles independently of one another. Although this should be substantially eased by the use of v4 profiles, in which the PCS reference medium to be assumed in perceptual profiles is more precisely defined than previously, the wide differences in gamut and media which may be encountered between input and output, as well as the effect of image content, place a significant difficulty in the path of a vendor or user optimizing such profiles. While non-colorimetric renderings in profiles can be, and often are, optimized separately, the reproduction requirements of some high-quality market sectors require that final optimization can only be done for the pair of profiles to be employed in generating the color transformation.

Because of this, there are certain market sectors, notably in printing and publishing, where users prefer to optimize their profiles and convert to the output space (CMYK, possibly with additional separations) early in the process, or even use proprietary methods for producing the separation data. This introduces an issue for some users where they wish to exchange separation data, but later use profiles to convert to other output device spaces which use the same number of colorants. Because ICC profiles use the PCS as the reference, it means that, while a color match should be maintained, the relationship between the separation values for each pixel is most likely to change. For many users this is often unimportant for many images (as, for example, they select the GCR they require for their own printing conditions), but not for files where elements are defined in only one or two separations (such as black-only borders and text). For such elements, maintaining the separation composition can often be more important than precisely matching the color.

Increasingly this will change for many users where the final output profile will be used in proofing simulations, but without conversion of the data until it is finally output. This will necessitate the exchange of both input and output profiles, but file formats should be extended to permit this, as PDF/X already has. However, that will not suit every workflow and there will still be a requirement in many workflows to convert separation data. In such situations it is important for some users to be able to retain a good approximation to the K to CMY ratios. Such algorithms are not particularly complex and are provided by a number of software vendors; however, there has been no agreement within the ICC on any particular method. The ICC is considering methods for specifying CMYK to CMYK conversions, but given the lack of definition as to exactly what trade-off between maintaining separation and color accuracy is required, beyond the obvious requirements, it is not going to be easy to get agreement. Many users employ software that provides the functionality required for their particular needs, and use it to provide DeviceLink profiles. Others use "smart" CMMs that offer this functionality.

ICC DeviceLink profiles provide a means to encapsulate a transform from one set of device values to another in a standard format. However, while DeviceLink profiles could be considered to maximize predictability while minimizing flexibility, they only provide one transformation intended for a specific pair of devices. "Smart" CMM functionality has the potential to offer the most flexibility, and therefore the least predictability. Typical ICC color management using two profiles is somewhere in between, because of the fixed transforms but different rendering intent options.

3.3.4 Device Calibration

In order for any color reproduction system to produce satisfactory results it is necessary that all the devices in the system behave as they did when the color transformations employed in the system were established. This is the case for both proprietary and ICC-based systems. In the ICC context it means that a profile is only valid for the state of the device, and the media used, at the time it was made. However, many devices are not inherently stable. So, in order to reduce the need to make a new profile every time a device changes, some mechanism for bringing the device back to the state it was in when a profile was made is highly desirable. However, the mechanism used for correction will vary with the device – sometimes physical changes are made to the device itself; at other times changes are implemented in the control software. Sometimes calibration requires measurement of control elements, and input of the resultant data, by the user; at other times it is an automatic procedure.

Whatever procedure is involved it must be the responsibility of the user to ensure that any devices used by them are maintained at a level consistent enough to ensure that the profiles for those devices remain valid, and if a device deviates beyond the levels which allow satisfactory use of existing profiles, a new profile will need to be produced.

Because of the wide range of calibration procedures which may be encountered, and the differing deviations which may be acceptable to any market sector, it is not possible to make definitive recommendations concerning calibration. A profile can optionally contain tags that provide calibration data describing the status of the device at the time the profile was made. However, even if these tags are filled, and used by color management software to compare to data measured at the time of the calibration, it is the responsibility of the software vendor, or user, to provide the acceptable tolerances beyond which correction needs to be undertaken. Nevertheless, once this is done the tag information can be used by the color management software to initiate the necessary correction procedures. To ensure good-quality reproduction, it must be the responsibility of the user to properly calibrate the devices used – where this is not satisfactorily achieved automatically – either directly, or within the color management software where that functionality is provided.

3.3.5 Measurement Procedures

The CIE system of color measurement has a number of variants. In order to avoid ambiguity it is necessary for any color reproduction system defined with reference to CIE colorimetry to be precise about the measurement conditions to be used. This has been achieved by the ICC by requiring that measurement of reflecting and transmitting media be consistent with ISO 13655. The requirements that this imposes, together with recommendations for the measurement of displays, are summarized in Chapter 20 below.

However, although by far the majority of users will want to follow these recommendations, the ICC profile also contains a number of optional tags that enable users to specify alternative measurement conditions, observers, and viewing conditions, and a required tag to define the chromatic adaptation needed when predicting equivalent color for adaptation to illumination of a different chromaticity to that of D50. Since in most cases there is no well-defined color conversion between the various conditions specified (except where the tags are used to define the parameters needed by color appearance models), the ICC does not make specific recommendations about how these tags are used, but for market sectors where alternative conditions are important, they can be utilized by the profile creation software and CMMs using these profiles to provide the desired functionality. Similarly, characterization data (both spectral and tristimulus, if desired) can be included in the profile to enable color management software to utilize it where appropriate.

3.4 Summary

As should be clear from the above discussion, color transformations of considerable complexity can be defined by using ICC profiles. These can be used to define the transformation at each stage in the color reproduction system where a color conversion is required. While "basic" profiles will be adequate for many applications, they can be edited to provide particular requirements as necessary. In some market sectors there may be requirements that go beyond the functionality offered by the basic profile approach – such as CMYK to CMYK conversions that retain the K. Vendors of color management software serving these markets usually provide this functionality, often in the form of DeviceLink profiles.

For certain market sectors, where additional information can be utilized by the color management software, various optional tags can be filled with the information necessary to provide additional functionality. It would be impractical for the ICC to define what is mandatory in such cases, as for many markets keeping the profile size to a minimum is of utmost importance. However, in industry sectors where such optional information would be useful, "standards" groups serving those market sectors should be encouraged by their users to get agreement among vendors in that sector as to what should be mandated for those applications.

Many captured digital images have characteristics that would not be acceptable if reproduced colorimetrically on a particular medium which is different to that assumed in creating the original image. In order to achieve satisfactory reproduction such images need to be color re-rendered. Perceptual intent transforms are intended to achieve this automatically; however, users need to ensure that such transforms are satisfactory for their needs and apply further editing if necessary.

4

Common Color Management Workflows and Rendering Intent Usage

The ICC architecture supports the flexible definition of color management transforms and workflows that can be used for many different purposes. A wide variety of color reproduction goals can be implemented through the choice of appropriate ICC profiles and rendering intents. However, this flexibility can cause confusion, so it is therefore useful to document some common of the most common workflows and provide guidance on rendering intent selection and usage.

It is also important to distinguish between workflows in which the two main versions of the ICC specification (v2 and v4) are used, because a number of limitations and ambiguities in the v2 specification (ICC.1:1998–09 and earlier) were removed from the v4 specification (ICC.1:2001–12 and later). Further clarification has been provided with the publication of the updated 4.3 specification (ICC.1:2010).

4.1 Common Color Reproduction Goals

Two common objectives in color reproduction are re-purposing and re-targeting. Re-purposing is performed when color content that has been color rendered so that it is optimal for one output color encoding is subsequently color re-rendered to make it optimal for another output color encoding. The typical display and printing of photographic images involve re-purposing because each picture is initially color rendered to an intermediate color encoding (such as sRGB or the ICC PCS perceptual intent reference medium), and then is subsequently color re-rendered as needed for a specific display or print output. With re-purposing, the objective is to take an original that is assumed to be optimized for the particular medium and viewing conditions (e.g., the ICC PCS perceptual intent reference medium), and create a new, optimal

Color Management: Understanding and Using ICC Profiles   Edited by Phil Green
© 2010 John Wiley & Sons, Ltd

reproduction on a second medium, possibly with different viewing conditions. While there may be a natural desire to maintain some level of consistency between original and reproduction, if the media are different there will likely be intentional differences in the colorimetry. This is because the objective is to make the "best possible" reproduction in each case, which will depend on the reproduction media and on the particular preferences that may be applicable to the media and the content.

It is important to note that, with re-purposing, selection of the best possible reproduction will be subjective, and will depend on viewer preferences. This means it is not possible to standardize re-purposing transforms. They will remain proprietary, and different transform creators may intentionally produce different transforms for the same original and reproduction medium combinations, to address different user preferences. The success of any particular transform will depend on whether users like the results of using it. While it is reasonable that users will want some level of consistency with the original, it is unlikely that exact colorimetric reproduction will be preferred when there are significant differences between the original and reproduction media or their viewing conditions.

Re-targeting can be thought of as an alternative to re-purposing. Re-targeting is performed, for example, when a proof is made that is intended to match a reproduction on different media. Re-targeting is distinct from re-purposing because the reproduction goal is to produce not the best reproduction possible, but rather the closest possible match to some other target reproduction. Re-targeting may include colorimetric adaptation, when the white bases of the proofing and original media are different and adjustment is necessary to compensate for the different state of adaptation when the two media are viewed sequentially; or it may ignore such differences between the media whites and simply aim to match the colorimetry of the target when both media are measured relative to a perfect reflecting diffuser. Regardless of whether such compensation is applied, the intent is always to preserve the original rather than to reshape it in some way.

Conceptually, the "perceptual" and "saturation" rendering intents are intended for re-purposing and the colorimetric rendering intents are intended for re-targeting operations such as proofing. There are two specified colorimetric ICC rendering intents, media-relative colorimetric and ICC-absolute colorimetric, with the ICC-absolute colorimetric intended for cases in which the proof is desired to include the look of the original medium. Note that in practical workflows there is some overlap and in between area with the use of these rendering intents. For example, with ICC v4, if a target medium is very similar to the perceptual intent PCS reference medium, the perceptual and colorimetric rendering intent transforms in an output profile targeting that medium may be identical – reflecting the fact that in this case color re-rendering is not required and only a re-encoding transform is needed. Similarly, when "original" and target output media and conditions are similar, the media-relative colorimetric rendering intent can be combined with black point scaling as a minimal perceptual rendering intent, useful for re-purposing between the two output representations.

With ICC v2, the choice of rendering intents is sometimes dependent on issues which arise from limitations of the v2 specification, instead of being solely based on the reproduction goal. For example, a v2 output profile perceptual transform may be specifically designed to receive black-scaled sRGB colorimetry in the PCS and re-render it to an inkjet photo medium. However, if an original is a photographic print that has been scanned colorimetrically, it would not be appropriate to apply a color re-rendering that assumes sRGB as the source. With this output profile, a better choice of rendering intent for the photographic print original would be to use the media-relative colorimetric intent with black point scaling. Re-purposing is still desired, but the media-relative colorimetric with black point scaling is more appropriate for color re-rendering the photographic print original than a perceptual intent that assumes an sRGB original. On the other hand, if such a profile contained effectively a "media-relative colorimetric with black point scaling transform" in the perceptual intent, it would then not be able to produce as good results with sRGB originals. Note that one aspect of complexity with v2 in particular arises from these "special case" profiles.

If one wants to proof an original, the ICC-absolute intent will provide the most accurate colorimetric reproduction, particularly when the proof is intended to represent the color characteristics of the target medium. Note, however, that using the v2 ICC-absolute colorimetric rendering intent can be problematic, because of ambiguities in the v2 specification about whether the media white point recorded in the profile (and hence used to calculate the ICC-absolute colorimetric transform) is before or after adaptation to the D50 PCS adopted white has been performed. If a v2 profile with media white point that has not been chromatically adapted to D50 is combined with a profile with a D50-adapted media white point, an inappropriate color cast will result. There are also issues if the proofing media white is very different from the original media white. While it is not advisable to use proofing media with a white that is very different from that of the original, small differences can be accommodated using the media-relative colorimetric intent. The use of this intent also avoids the adapted/non-adapted media white point ambiguity because the colors are referenced to the media white. Note, however, that the media-relative colorimetric proof will not accurately represent the original media.

It is important to note also that ICC color management always assumes that the original image content is exactly as desired for its medium. Image correction is outside the scope of ICC color management. However, ICC profiles can be and are sometimes used in closed workflows to correct images with specific flaws.

Care should be taken to manage such profiles, because they can cause problems if they are used within open color management workflows. While this is an example of the flexibility of the ICC profile format, care must be exercised in incorporating such specialized elements into workflows in order to maintain interoperability.

The use of such image-specific profiles can combine both initial color rendering of scene-referred image data with image correction (without the need to change the original image data) into a single operation, but must be understood to be specific to certain images or sets of images, and managed accordingly. When image-specific profiles are used, the PCS description provided by the profile effectively becomes the "original" which is to be matched in subsequent color management operations.

4.2 Profile Functions

ICC profiles (both v2 and v4) perform two functions. The first, coordinate transformation, relates device color code values to colorimetric code values in the PCS. The second, color rendering or color re-rendering, changes the colorimetry of an original to be better suited for some particular reproduction medium. These functions are distinctly different, and each may or may not occur in a given transform in a profile. When they both occur within a particular profile/rendering intent, they are folded together in the particular profile/rendering intent transforms. This combining of color reproduction goals can be a source of confusion, so it is helpful to clearly distinguish between the coordinate re-encoding transforms necessary to convert back and forth between device values and colorimetry, and the color rendering and re-rendering transforms that alter colorimetry to achieve specific reproduction goals. The different rendering intents in a given profile express these different color rendering and re-rendering transforms.

Coordinate transforms can be determined objectively using measurements and characterization models, and as a result there is little debate on how they should be constructed (although there have been issues about how to take measurements and correct for flare appropriately when different media require different measurement devices, which are addressed in Chapters 10 and 20). Consequently, the remainder of this chapter will focus on color rendering and re-rendering, and it will be assumed that appropriately determined coordinate transforms are incorporated with the color rendering and re-rendering transforms as needed.

4.3 Profile vs. CMM Rendering Intents

It is helpful to distinguish between rendering intents in which the transform applied is explicitly included in the profile (media-relative colorimetric, perceptual, and saturation) and the ICC-absolute colorimetric rendering intent transform that is calculated by the CMM from the media-relative colorimetric intent contained in the profile. In a sense, the ICC-absolute colorimetric rendering intent could be considered as the first implementation of a "smart" CMM, in which the profiles contain the coordinate transforms between device values and colorimetry, and the complete transform is computed by the CMM from the profile data.

Version 4 profiles can support a range of smart CMM functionality, where the rendering intents are computed in the CMM at run-time, because the colorimetric intents are unambiguously defined and are required to be based on standard measurements. Further details are given in other chapters.

4.4 ICC v2 Rendering Intents

The common v2 ICC workflow involves the use of an input profile with a single (usually unidentified) rendering intent. This rendering intent will typically be based on one of the following:

- A media-relative colorimetric rendering, which transforms the input device values into media-relative colorimetry of the original in the PCS (coordinate transform relative to media white). A typical example would be a scanner profile obtained using an IT8 scanner characterization target.
- A media-relative colorimetric rendering, but with black point scaling where the black point of the original medium is scaled to zero in the PCS (coordinate transform plus black point scaling). A typical example is the v2 sRGB profile.

- A perceptual rendering, where the input-side transform color renders (e.g., a digital camera profile) or color re-renders (e.g., a "tuned" transparency scanner input profile) the colorimetry of a scene or an original to some proprietary virtual medium in the PCS (coordinate transform plus proprietary color re-rendering). These profiles commonly result from modification of one of the above transforms "to make the results better." It should also be noted that, with v2 perceptual intents, the proprietary virtual medium black point is scaled to zero in the PCS. Following the input profile, a v2 output profile is used to color re-render the colorimetry in the PCS to the output medium, and create output device values. Output profiles (whether v2 or v4) can have ICC-absolute colorimetric, media-relative colorimetric, perceptual, and saturation rendering intents, which are identified in the profile and selected to achieve the desired reproduction goal.

Ideally, the v2 output profile perceptual intent will be constructed to receive the colorimetry as transformed to the PCS by the input profile. This means that the perceptual intents of different v2 output profiles should be matched to specific input profile behavior. In this sense, the ICC v2 specification is more of a standard format for color profiles (as the title of the specification indicates), as opposed to an interoperable color management system. However, the user need is for an interoperable color management system and to some extent rendering intent selection can be used to achieve this.

For example, one might think that v2 colorimetric intents should be used for proofing, and v2 perceptual intents should be used for re-purposing. This is true for matched profiles, but what if there is a need to use an output profile with different types of inputs, and matching input profiles are not available? The provider of the original might not know which output profiles are to be used, or might not have the ability to construct matching input profiles. In this case, it is advisable to select the output-side rendering intent that best accommodates the PCS colorimetry produced by the input-side profile. For example, if one is reproducing a scanned photograph on a print medium, it may be acceptable to use media-relative colorimetry for re-purposing, especially if CMM black point scaling is applied. This is because the media-relative and black point-scaled colorimetry does a reasonable job of color re-rendering. However, if the original is an sRGB image, a more elaborate perceptual intent transform may be needed to re-render the black-scaled sRGB display gamut to that of the print medium.

In this example, the media-relative colorimetric intent with black point scaling is used for re-purposing from one print medium to another, and the perceptual intent is specifically designed to accept sRGB image data and re-purpose it for print. This allows a single v2 output profile to support two different types of input.

The problems with the above solution are that it complicates rendering intent selection, and that a number of rendering intents or profiles are required (at the limit, one for each type of medium for which the image data in the PCS is appropriate). The user needs to know what the output profile perceptual intent was constructed to receive, and what the input profile is putting in the PCS. There also needs to be an output-side rendering intent available that is designed to receive the data that the input profile supplies.

When choosing the output profile rendering intent with v2 profiles, it is advisable to consider both the reproduction goal and the colorimetry represented in the PCS by the input profile. This situation frequently leads to user frustration, since many users do not have the means or knowledge to analyze what different profiles are doing, other than by viewing the results of different combinations.

4.5 ICC v4 Rendering Intents

The ICC v4 solution is preferred, since the selection of v4 rendering intent is greatly simplified by clearer definitions of the rendering intents and the associated requirements for what the rendering intents contain.

Some v4 clarifications are as follows:

- The "relative colorimetric" rendering intent is renamed "media-relative colorimetric," to make it clear that it is media white relative. It is also required to be based on 45:0 measurements made according to ISO 13655 for reflecting media and appropriate measurement techniques for other media (as described in Chapter 20 below). It is therefore exclusively a coordinate transform.
- The "absolute colorimetric" rendering intent is renamed "ICC-absolute colorimetric," to avoid confusion with CIE terminology for colorimetry. The ICC-absolute colorimetric rendering intent is calculated by the CMM from the media-relative colorimetric rendering intent in the profile, and is a coordinate transform that places CIE colorimetry (relative to a perfect reflecting diffuser illuminated by a D50 illumination source) into the PCS.
- The media white point is required to be chromatically adapted to D50, so that v4 ICC-absolute colorimetric rendering intents will all be interoperable.
- A key benefit that derives from the strict v4 colorimetric intent definitions is that various smart CMM rendering intents can be computed using this measurement-based data.
- Input profiles can have multiple, well-identified rendering intents, with the same requirements as output-side profiles.
- The PCS print reference medium dynamic range and viewing conditions are defined for the perceptual intents to color re-render to and from. This enables perceptual intents to be interoperable (although if the re-rendering contained in a perceptual intent transform is poor, the results may also be poor). The reference medium gamut is recommended to be the print-referred gamut defined in Annex B of ISO 12640-3, known as the Perceptual Reference Medium Gamut; documentation of this gamut is provided in the ICC specification. Note that there can be interoperability problems if different perceptual intents re-render to and from very different PCS reference medium gamuts. However, one would expect that reference medium gamut assumptions for a print reference medium with the defined dynamic range would not be too different.

When the above clarifications are incorporated in v4 profiles, it is usually not necessary to consider the matching of the input- and output-side profiles when selecting v4 rendering intents. With v4 profiles, rendering intent selection is based on the reproduction goal.

4.6 Re-Purposing Using ICC v2

The method for selecting rendering intent using v2 profiles for re-purposing is as follows:

1. Determine what type of colorimetry the input-side profile will be placing in the PCS (e.g., sRGB display colorimetry, photographic print colorimetry, photographic transparency colorimetry, oil painting colorimetry, etc.). This determination involves the nature of the original, and also any input-side color re-rendering that the profile may be performing.

2. Locate an output profile rendering intent that performs the desired color re-rendering from the PCS colorimetry created by the input profile to the actual reproduction medium. This may involve the selection of both an output profile and a rendering intent transform within that profile. One may find an output profile where the perceptual intent is designed to receive the PCS colorimetry. If no suitable perceptual intent is available, a colorimetric intent (with or without black point scaling) can be tried to achieve a simple color re-rendering. As mentioned above, typical users may not have the capability to perform these two steps except by trial and error.
3. Use the output profile rendering intent that produces the desired result.

4.7 Re-Purposing Using ICC v4

Generally, the perceptual rendering intent should be used for re-purposing with v4 profiles. The exception is if the reproduction goal is to reproduce the PCS perceptual intent reference medium colorimetry, in which case the perceptual intent is used for the source transform and the media-relative colorimetric intent is used for the destination profile. In cases where the actual medium has a smaller color gamut than the perceptual intent reference medium, this practice will usually produce more colorful results, but at the expense of some clipping in the final image. Perceptual intent transforms typically attempt to maintain detail throughout the color gamut by applying compression or expansion as needed to re-purpose the perceptual intent reference medium image to the actual medium gamut.

It should be noted that most current color management applications and CMMs do not support selection of different rendering intents for source and destination. Also note that there is a saturation rendering intent, which remains relatively undefined with v4 profiles. It can contain color re-rendering transforms constructed to meet specific proprietary requirements, and can also be used for other purposes.

4.8 Re-Targeting (Proofing) Using ICC v2

Re-targeting using ICC v2 involves the following steps:

1. Locate input (source) and output profiles where the relative colorimetric intents contain accurate media white relative coordinate transforms only (no black point scaling; no subjective tweaking or color re-rendering). Ideally, these profiles should also contain media white points that are chromatically adapted to D50, to enable interoperability of the absolute colorimetric intent.
2. Determine whether the absolute or relative colorimetric intent produces the desired result. If the original and reproduction media whites are identical or similar enough, both rendering intents will produce the same result and it will not matter which is used. As the media whites diverge, a trade-off is encountered. The absolute colorimetric intent will tend to produce a better match of colors when individually compared (as is usually the case for spot colors, for example), while the relative colorimetric intent may produce a better overall appearance match (due to a degree of human visual system adaptation to the different media whites). It may also be possible with some CMMs to produce hybrid transforms, where, for example,

the original media white luminance is scaled to that of the reproduction media white, but the original media white chromaticity is left unchanged. This type of hybrid transform avoids highlight clipping (as does the media-relative colorimetric rendering intent) as well as chromaticity shifts.

4.9 Re-Targeting (Proofing) Using ICC v4

The same workflow applies here as for proofing using ICC v2, except the first step of selecting input and output profiles with appropriately constructed colorimetric intent transforms is unnecessary, since all valid v4 profiles meet these criteria.

5

Recent Developments in ICC Color Management

5.1 Introduction

In an ICC color-managed workflow, profiles constructed according to the ICC profile format standard [1] are used to transform between color encodings, as needed to produce the desired manifestations of an image as reproduced on displays and prints, and saved in digital files. In any particular transform, the starting color encoding is called the source encoding and the resulting color encoding is called the destination encoding. The source profile defines the transform from source color encoding into the profile connection space (PCS), while the destination profile defines the transform from PCS values to the destination color encoding. (Note that we use the term "color encoding" here rather than "color space" as we need to include the specification of image state [1], reference medium, and viewing environment as well as the color space itself.)

ICC profiles use a tagged format. A profile is made up of a 128-byte header, followed by a tag table and a series of individual tags. These tags can be informational or numeric and, depending on the profile class, may also be optional or required. The profile version (v2, v4) is located within the profile header, together with information about the intended conditions of use of the profile. An introduction to the ICC profile is published by the ICC [2].

Version 4.2 of the specification (ICC.1:2004–10 [3]) was published in late 2004. This version was approved by ISO TC 130 and published as ISO 15076-1 in 2005 [4]. The ISO publication incorporated minor errata, some of which had already been published by the ICC, which then republished Version 4.2 to incorporate the errata, resulting in the ICC version available being identical to ISO 15076-1. Since Version 4.2, the ICC has approved a number of amendments, and these are reviewed in more detail below. These amendments are incorporated in the v4.3 (ISO 15076-1:2010) specification, together with some further rewriting of the specification to reduce ambiguity.

Color Management: Understanding and Using ICC Profiles   Edited by Phil Green
© 2010 John Wiley & Sons, Ltd

5.2 Changes in v4

The ICC v4 major revision was necessary because it had become apparent that the v2 specification [3] gave rise to a number of problems. The main issue was that color management could lead to different results depending on the particular vendor's interpretation of the specification. Interoperability problems arose as profiles from different sources made different assumptions and sometimes did not work well together. To make matters worse, CMMs (Color Management Modules) could produce different results depending on how they dealt with inconsistencies in the profiles.

Experience showed that there were issues with the colorimetric and perceptual rendering intents. In the colorimetric intents, the mediaWhitePoint tags were sometimes incorrectly encoded, that is, without applying an appropriate chromatic adaptation transform. This led to the ICC-absolute colorimetric intent values being incorrectly calculated. Some CMMs chose to resolve this issue for display profiles by assuming full adaptation to the display white point and ignoring the media white point in the profile. While this fixed the problem with bad profiles, it also led to incorrect interpretation of good profiles in which the profile creator had chosen to assume no or partial chromatic adaptation. In some cases problems resulted with correctly constructed v2 profiles, because different chromatic adaptation methods were used, but not specified in the profile, making it impossible for the CMM to get back to the measured colorimetry and apply a single method. Also, some vendors did not provide true colorimetric transforms but incorporated an element of re-rendering into the colorimetric intents. For example, black point scaling is common in v2 display profiles. Because the v2 colorimetric intent PCS colorimetry cannot be relied on and it is not possible to get back to the original, unadapted source or destination colorimetry, the colorimetric transforms in v2 profiles provide poor support for the smart CMM functions that are increasingly being used.

There was a particular problem with the perceptual rendering intent, where the specification was not clear on the PCS reference medium. As a result, output profiles made very different assumptions about the PCS reference medium. At one extreme, a profile might attempt to render the whole of the PCS encoding range to the destination gamut, with perhaps a scaling of the black point, while another profile might assume a much more limited PCS reference medium gamut such as that of a CRT display. This potentially leads to significantly different results even for similar media.

In the absence of a well-defined reference medium, an ad hoc practice evolved in which black point scaling was applied in perceptual transforms and sRGB was assumed as the perceptual reference medium. This generally gave acceptable results when the source media were similar to a display in color gamut. For other source gamuts, media-relative colorimetric rendering intent with black point compensation was adopted. However, this ad hoc approach gave rise to other problems when using the perceptual rendering intent. It meant that selection of rendering intent depends on the source profile as well as the color rendering goal, which makes both manual and automated workflows more complicated. It also tends to produce poor results when the perceptual intent is used and the source is not display-like. And since this ad hoc workflow is not standardized or even documented, users cannot assume that all profiles have been constructed in this way.

In the v4 revision process, there was a thorough review of the specification to remove virtually all of the ambiguities [5]. New requirements that were adopted in v4 included the following:

- Colorimetric rendering intents are now based directly on the measurements from which the profiles were generated, without the "tweaks" that were seen in some profiles, and any black point scaling is to be performed by the CMM.
- It has been clarified that the values to be encoded in the mediaWhitePointTag should be after chromatic adaptation to the D50 PCS, so that all PCS-side data in a profile is D50.
- When the source colorimetry is not relative to a D50 adopted white (as is often the case with display profiles), v4 introduced the requirement to specify the chromatic adaptation matrix so that the CMM can "undo" the transform if necessary.
- For *n*-color profiles, the CIELAB values of all the device colorants must be specified.
- A Perceptual Reference Medium (PRM) is defined in v4. This PRM has a reference viewing condition based on the P2 viewing condition for appraisal of hard copy prints in ISO 3664 [6], with an illuminance of 500 lux and a neutral surround reflectance of 20%. Whereas the v2 PCS has a reference medium based on an ideal print, defined as a perfect diffuse reflector with an unlimited color gamut, the PRM is based on a high-quality virtual photo print with a 288:1 dynamic range, having a neutral reflectance of 89% and a darkest printable color having a neutral reflectance of 0.309 11%.
- The perceptual rendering intent is required to color-render or re-render to and from the specified standard PRM.
- Multiple rendering intents can now be included in all table-based profiles, including input and display profiles. For CRT displays, the simple matrix-based type of profile was generally adequate. With the move to other display technologies, the assumptions of a matrix-based profile does not hold, and a look-up table (LUT) will often give better results.
- Version 4 profiles can include matrix-based transforms in addition to LUTs. This applies to input profiles as well as output profiles.
- New extended tag types for color LUTs were introduced, together with a more flexible parametric specification for curves.

Two other additions to the v4 specification are as follows:

- Unicode was added for text types to assist in providing multi-language versions of profiles.
- A profile ID (a more or less unique identifier for each profile, calculated using the MD5 hash algorithm) was added to the header; this is useful in checking which profile is used, especially in cases where two profiles have the same name but different content.

As well as adding new tags, it had been observed that some v2 tags were in practice almost never used and it was agreed to simplify the specification by removing them. Finally, the specification was restructured and informative material expanded and rewritten to improve the clarity of the document.

5.3 Perceptual Rendering

A perceptual rendering is one where the goal is to produce a pleasing reproduction of an original (the source) on some destination output medium. This is also known as a "preferred reproduction." Note that the reproduction in this case is not required to be an exact match to the original, although if the original image state is output referred (i.e., the data represents a pleasing reproduction of the image on some output medium), then the artistic intent of the rendering to the output-referred original should be maintained.

The perceptual intent of ICC profiles can also be used to color-render scene-referred images (i.e., images where the data represents the original scene colorimetry, not an interpretation of the image rendered to an output medium). Preferred (or "perceptual") reproduction transforms address differences in source and destination media capabilities and user preferences, as well as viewing conditions.

The v2 format limited the ability of the profile to provide suitable transforms between different image states. Different vendors had different interpretations of the transforms between source data and the PCS, particularly affecting the perceptual rendering intent.

Although the v2 ICC PCS is D50 output referred, most source profiles performed minimal color rendering or re-rendering to the PCS (usually only black point compensation (BPC)). On the output side, profiles had no information on the source color gamut, and vendors made different assumptions about the source reference medium and gamut when color re-rendering to the output medium.

With v4, the source profile perceptual intent transform is intended to color-render or re-render to the v4 PRM, and similarly the destination profile transforms from the PRM to the destination medium.

The reflection print character, dynamic range, and viewing conditions of the PRM were defined in the v4 specification, but not the target gamut. The PRMG amendment [7], the first amendment to the ICC.1:2004–10 specification, adopted the Perceptual Reference Medium Gamut (PRMG) as the target gamut, thus for the first time providing a necessary and sufficient definition of the PRM. (It should be noted that the PRMG is intended as a "fuzzy" gamut, and it not necessary to color-render or re-render *exactly* to and from this gamut.) The PRMG amendment provides optional tags for the perceptual and saturation rendering intents that identify whether the PRMG was used in rendering to or from the PRM. The PRMG is illustrated in Figure 5.1.

Although intended to represent the gamut of the PRM (a high-quality photographic print medium), the PRMG also approximates the gamut of commonly occurring surface colors (although a few reproducible colors will be found to lie outside the PRMG). A full specification, including data on the gamut surface and primaries, is given in ISO 12640-3 [8].

Most input (camera, scanner) and color space profiles in circulation are v2 profiles, and do not have perceptual intents that render to the v4 PRM. The ICC has addressed this for profiles which render from the sRGB standard color encoding by recommending a v4 sRGB profile whose perceptual intent re-renders from the sRGB reference display to the PRM, including the PRMG. This profile is available for download from the ICC web site. The ICC recommends that vendors develop similar rendering/re-rendering capability in their v4 input and color space profiles.

Similarly, most existing printer profiles are v2 profiles, or, if they are v4 profiles, they do not have the PRMG as the assumed source gamut for the perceptual intent. The ICC recommends that vendors use the PRMG as the source PRM gamut in their v4 output profiles.

Fundamentally, a profile identified as a v4 profile requires these improvements in transform behavior, in addition to the various format changes described.

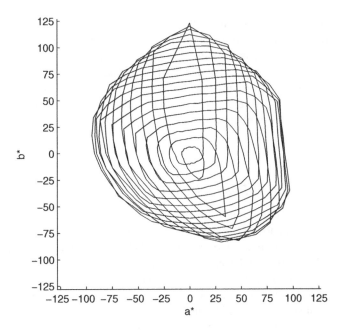

Figure 5.1 Perceptual Reference Medium Gamut at CIELAB L^* 5–95 in increments of 5 in L^* (ICC media-relative colorimetry)

5.4 Other Developments Following ICC.1 2004–10

5.4.1 The Profile Sequence Identifier Tag

The DeviceLink profile class has been part of the specification for some time, and is important in many graphic arts workflows, particularly where the user needs to preserve the channel intent of an original. For example, if a graphic or text is specified in CMYK amounts, it is often undesirable to convert to the PCS and then back to CMYK using the final destination profile, as the balance between the colorant amounts (also known as the "channel intent") may change.

A DeviceLink profile LUT is normally generated from the AToB and BToA LUTs present in a pair of profiles used as source and destination, although a DeviceLink profile can also be constructed from a series of such profiles which together form a processing pipeline.

One limitation of DeviceLink profiles was that they contain insufficient information on the profiles used to construct the link. The profile sequence identifier tag [9] allows this information to be recorded within the profile.

5.4.2 Profile Registry

The Profile Registry [10] is a new resource for color management users. It provides a publicly accessible repository for profiles which correspond to registered printing conditions, and it supports both manual and automated location of profiles.

Manual selection is supported through a list of registered profiles and the printing conditions they correspond to. Details of the printing process and the organization providing the profile are available, together with further information for each registered profile.

Automated location of profiles is needed by the proposed ISO 15930 [11] PDF/X-4 and PDF/X-5, which allow users to include links to a resource rather than including the resource itself within the PDF file. Not having to include the profile within each file reduces overall file size and avoids duplication, and this is particularly important in variable data workflows.

5.4.3 Colorimetric Intent Image State Tag

Most ambiguities with the colorimetric and perceptual rendering intents have been resolved in the v4 specification and the PRMG amendment. When image data is transformed perceptually to the v4 PCS, it is clearly output referred, with the PRM as the reference medium. However, when using the colorimetric rendering intents, the PCS colorimetry is required to be based on measurements, so the image state will depend on what was measured. Consequently, the ICC has added a Colorimetric Intent Image State tag [12] to the specification, which allows the image state to be specified for the colorimetric rendering intents. When this tag is not present it is assumed that the colorimetric intent PCS colorimetry is "picture referred" – that is, it represents a hard copy or soft copy reproduction rather than original scene colorimetry.

The scene-referred signatures are scene colorimetry estimate, scene appearance estimate, and focal plane colorimetry estimate. In addition, signatures are available for photographic hard copy originals and print-referred output. The effect of this tag is to enable a user to correctly interpret the PCS values which result from applying the colorimetric rendering intents to the source image data.

If an image has a Colorimetric Intent Image State (CIIS) tag with a signature of "scoe" or scene colorimetry estimate, then the colorimetric intents can be used to convert the image to a scene-referred color encoding (or working space). If a colorimetric reproduction of the original scene were desired, then the ICC-absolute colorimetric rendering intent would be used for this conversion. Alternatively, if some form of manual color rendering is to be performed, the media-relative colorimetric rendering intent will map camera saturation white to the scene-referred working space white point, to avoid clipping. The perceptual intent could also be used to convert the image to an output-referred color encoding, for screen viewing or to make a print.

5.4.4 Floating Point Device Encoding Range Tag

Color management is becoming increasingly important as the motion picture industry goes digital. The typical color reproduction workflow involves capturing digital data by scanning camera negatives or using digital cameras, converting source data to a working space (which is often scene referred) for editing and compositing, and then to either the DPX format for writing the final output negative for film printing or to the Digital Cinema Distribution Master encoding for digital theatre presentation. In some cases, the digital motion picture workflow requires that transforms can be perfectly inverted. The particular form of the transforms, particularly tone reproduction curves, means that transforms cannot be performed precisely within the precision of the integer processing elements in the ICC profile format, and quantization errors can result.

A second issue for motion picture color transforms is that in the ICC specification device data is bound to some specified media white, while motion picture workflows commonly incorporate scene data in unbounded floating point encodings. The Floating Point Device Encoding Range amendment [13] addresses these limitations by providing a new, optional set of transforms, which are known as DToB and BToD transforms. The data type for these transforms is a signed, 32-bit floating point encoding.

Associated with this new type of transform is the option to specify a sequence of processing elements, instead of using a fixed sequence as in the current LUT types.

The processing elements permitted by the amendment can be any sequence of matrix, curve, and CLUT. There is also some discussion about allowing these elements to connect profiles at places other than the PCS to provide a more flexible and extensible color processing architecture.

5.4.5 Smart CMM

A selectable sequence of multi-processing elements can be used by a smart CMM.

In a static workflow, which is most common today, all the adjustments required for viewing conditions, gamut mapping, and preferred renderings are in the profile itself, and the CMM merely performs the conversion specified within the processing elements of the profile.

The basic concept of the smart CMM is that additional transforms can be performed by the CMM, depending on the properties of the source and destination device and media. This allows optimization of the transform for the particular gamuts and viewing conditions involved, and also facilitates gamut mapping based on color appearance correlates. A smart CMM may also allow user selection or adjustment of transforms at run-time.

Rather than a simple dichotomy between static and smart CMM workflows, there is in fact a continuum between the static workflow on the one hand, in which all the smarts are in the profile, and the fully programmable workflow with dynamically configurable transforms on the other, in which the CMM or expert user selects which transform elements to use. ICC v4 profiles with a selectable sequence of multi-processing elements are applicable at any point along this continuum. Examples of the operations which can be supported by a smart CMM include: function inversion, black point compensation, selectable gamut mapping, color appearance modeling, adopted white scaling, adjustable chromatic adaptation, black generation, channel preservation, and any spatially varying or other proprietary operations such as scene relighting and color rendering.

5.5 The ICC.1:2009 Specification

Although the v4 specification resolved most of the significant ambiguities in the specification, it was considered that further editing was required to improve the clarity of the document and incorporate the amendments discussed above. Some of the edits were necessary as part of the process of transforming the document into an ISO specification, while others emerged from feedback from those implementing the specification. Minor errors were also found in certain specified parameters.

The ICC Specification Editing Working Group prepared a new version of the specification, which was approved by the ICC and by ISO TC 130 in 2009/2010. This version is known as ICC.1:2010, and profiles made according to this specification are Version 4.3.

In addition to the incorporation of the amendments, much of the explanatory material was rewritten to provide greater clarity and more consistent use of terminology. Most of the figures were redrawn for greater clarity. A clear distinction was made between CIE values, normalized CIE values, and the PCSXYZ and PCSLAB encodings.

5.6 Conclusions

The ICC v4 specification has established an architecture for interoperable and unambiguous communication of color between devices. As we look into the future, we can recognize that there are a number of challenges for ICC color management. Perhaps the most immediate challenge is to promote wider adoption of the current v4 specification. Following this, the context for color management is extending well beyond the traditional print-centric graphic arts. While many of the recent v4 amendments have been directed at extending the architecture to support the workflows of new types of devices and media, there may be further work to do in order to make the ICC profile support color conversion across a wide range of workflows.

The addition of multi-processing elements to the specification significantly extends the range of color processing models which are supported by the ICC architecture.

Profile creators and developers of color management applications should find that the ICC.1:2010 specification provides a clear guide to the writing, interpretation, and use of ICC profiles.

References

[1] ISO (2004) 22028-1. *Photography and graphic technology – Extended colour encodings for digital image storage, manipulation and interchange – Part 1: Architecture and requirements*. International Organization for Standardization, Geneva.

[2] Introduction to the ICC profile format. http://www.color.org/iccprofile.html.

[3] ICC (2004) 1:2004–10. *Image technology color management – Architecture, profile format, and data structure*. International Color Consortium, Reston, VA (with errata, 2006).

[4] ISO (2005) 15076-1:2005. *Image technology color management – Architecture, profile format, and data structure – Part 1: Based on ICC.1:2004-10*. International Organization for Standardization, Geneva.

[5] ICC (2003) 1:2003–09. *Image technology color management – Architecture, profile format, and data structure*. International Color Consortium, Reston, VA.

[6] ISO (2000) 3664:2000. *Viewing conditions – Graphic technology and photography*. International Organization for Standardization, Geneva.

[7] ICC (2005) Perceptual Intent Reference Medium Color Gamut Proposal. International Color Consortium, Reston, VA.

[8] ISO (2007) 12640-3:2007. *Graphic technology – Prepress digital data exchange – Part 3: CIELAB standard colour image data (CIELAB/SCID).* International Organization for Standardization, Geneva.

[9] ICC (2006) Profile Sequence Identifier Tag Proposal. International Color Consortium, Reston, VA.

[10] ICC (2006) Profile Registry Proposal. International Color Consortium, Reston, VA.

[11] ICC (2006) Colorimetric Intent Image State Tag Proposal. International Color Consortium, Reston, VA.

[12] ICC (2006) New Technology Signatures Proposal. International Color Consortium, Reston, VA.

[13] ICC (2006) Floating-Point Device Encoding Range Proposal. International Color Consortium, Reston, VA.

6

Color Management Implementation Classification

6.1 Overview

Color management is used and implemented in many ways. As different implementations and specific architectures are proposed it is useful to have a common conceptual framework within which these can be compared. This chapter briefly provides a definition of color management that can be used in the analysis of different architectural implementations. It then presents a general high-level architecture for color management and outlines a continuum for comparing different architectural implementations. In conclusion, different categories of architectural implementations are identified and compared using the presented continuum.

An important point to note is that there is no universal best way to implement color management. Each implementation will have its trade-offs as it achieves its goals related to color management, and the choices involved in these trade-offs are often different for different use cases. This chapter is intended to facilitate analysis and comparison of architectural implementations, and as such does not focus on specific workflows.

6.2 Color Management

The ICC glossary defines "color management" as follows.

6.2.1 Color Management (Digital Imaging)

communication of the associated *data* required for unambiguous interpretation of color content data, and *application* of color data conversions, as required, to produce the *intended* reproductions [ICC.1].

NOTE 1 Color content may consist of text, line art, graphics, and pictorial images, in raster or vector form, all of which may be color managed.

Color Management: Understanding and Using ICC Profiles   Edited by Phil Green
© 2010 John Wiley & Sons, Ltd

NOTE 2 Color management considers the characteristics of input and output devices in determining color data conversions for these devices.
(Italics added for emphasis)

Implementations of color management involve how four important parts from this definition are achieved: *communication, data, application, and intended reproductions.*

6.3 Architectural Layers of ICC Color Management

Generally, the architecture for current ICC color management is implemented in layers as shown in Figure 6.1. (Other architectures may exist but these layers can be thought to exist on a conceptual level.)

Generally the higher the level in the architecture shown in Figure 6.1, the more proprietary the implementation is considered to be. The lower levels are often considered to be more open.

Even though metadata in the lowest levels can be created using proprietary transform generation implementations, it is typically encoded in standard formats that can be used by more open implementations.

The top layer or application/driver layer is the client for color management. It ingests source image data and exports destination image data, possibly requesting lower levels to perform color management of the image data. This layer may gather and/or process color metadata, or may defer some or all gathering and processing to lower levels.

The color management system (CMS) layer processes color metadata, not color data. It obtains color metadata from the application level, from devices or their drivers, or from user input. The CMS determines the class of color metadata (such as OpenEXR CTL or ICC profiles), which in turn determines the class of CMM to use. In some cases, the color metadata can prescribe a preference for a CMM within its class.

The color management module layer assembles and executes color transforms. The CMM takes direction from upper and lower layers in addition to providing its own operational logic to perform transformations of the color data. Some CMMs can be used with only one class of color metadata, while other CMMs can be used with multiple classes. On some systems, multiple CMMs may be available for ICC profiles.

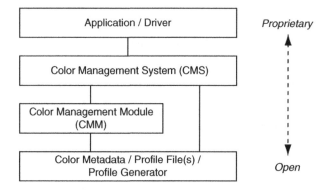

Figure 6.1   Color management layers

The lowest layer is the color metadata/profile layer, which provides information used to assemble and execute color transforms in the CMM layer. Color metadata may describe the characteristics of a color data source or destination, which are often related to physical or virtual reference devices/media. Color metadata may also provide color transforms and/or instructions for the application of color transforms. Many metadata formats are in current use. Some have variable digital representations, such as measurement data or transform data, while others are in the form of explicit or implied references to specifications (e.g., sRGB and the digital cinema $X'Y'Z'$). In the ICC workflow, the metadata is encoded as an ICC profile constructed according to the ICC profile format specification. Often, a metadata/ profile generator application is used to create the metadata/profile. Such generators can use their own operational logic in the process of generating the transforms encoded in the metadata/profile.

Since applications and/or drivers make all color management requests through the CMS layer, the term "color management system" often refers to the aggregate of the lower layers, instead of the top layer only. The context determines whether a single level or the aggregate is being referred to.

6.4 The CMM/Metadata Implementation Continuum

Most of the color transforms in a color management implementation are defined in the bottom two layers. The implementation possibilities can be considered as a continuum of run-time behavior with possible implementations of CMM and metadata layers at the extremes of each end. This can be seen in Figure 6.2.

If *the transform operations are defined and controlled well in advance of applying the color data transform(s)*, for example, when the color metadata defines the operations to perform, the implementation is classified as static. In this situation, the color metadata provides the complete operational logic, and the CMM needs no additional logic to determine what transforms to apply to the color data. This generally means that the operational logic in the transformations is assembled and used at the time when the color metadata is created. This is sometimes referred to as an early binding system.

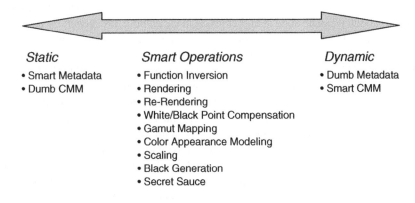

Figure 6.2   CMM/metadata implementation continuum

If *the transform operations are mostly defined by the CMM, user settings, and/or image data*, and not in the color metadata specification, the implementation is classified as dynamic. In this situation, the operational logic is provided by the CMM. Color metadata, if used at all, provides only basic color measurement information. A dynamic CMM is free to implement any operational logic that it wishes, but this comes at a cost of interoperability and predictability between dynamic CMMs with different implementations and/or configurations. This generally means that the operational logic in the transformations is assembled and used at the time when the transformations are applied. This is sometimes referred as a late binding system.

For most dynamic implementations, accurate color characterization data needs to be retrievable for the source and destination color data encodings. The ICC.1:2001–04 profile specification improved the ICC profile format to ensure that dynamic CMMs could retrieve accurate color characterization data from the profile.

It should be noted that current basic ICC implementations are not entirely static. Rendering intent linking, the XYZ to/from *Lab* conversion, and the absolute rendering intent operations to adjust the white point for ICC-absolute colorimetry represent dynamic run-time behavior required by the ICC profile specification. Additionally, CMMs that perform black point compensation also provide additional dynamic run-time behavior. Dynamic behavior is predictable when it is clearly specified in a standard. Thus the dynamic behavior is required to be available by the implementation and the specifics of when and how to apply the dynamic behavior are clearly defined.

6.5 Overcoming Limitations

With an understanding of the architectural layers of color management and the CMM/metadata implementation continuum, analysis of and comparison between different implementations are possible.

A basic ICC color management implementation, which supports only the transformations implied by the ICC profile specification, is limited to only those transforms that can be encoded in ICC profiles, or those that the CMM must dynamically implement as defined in the ICC profile specification. Additions need to be made somewhere in the color management layers to go beyond these limitations.

Changes made in lower layers of the architecture are easier to standardize for organizations like the ICC. Though it can be done at higher levels in the architecture, generally it is the CMM that is modified and possibly the color metadata. Different implementation approaches therefore correspond to movement in the CMM/metadata implementation continuum.

In a dynamic CMM implementation the sequence control is centralized in the CMM, but to be open and cross-platform, agreement on sequence/control within the CMM is required. In the past, reaching agreement has proven to be difficult. Some reasons include the significant preferential/artistic aspects of cross-media color reproduction, and the estimation of the color appearance of images viewed in different conditions. With such a lack of agreement, different CMM implementers have provided additional operational logic to address different use cases, possibly requiring private tags and/or external configurations to go beyond the limitations of basic ICC implementations. However, if private tags are used then they may not be understood by other CMMs. Interoperability between different dynamic CMMs is therefore limited to the baseline behavior required by the ICC profile specification.

6.6 Extending the CMM/Metadata Implementation Continuum

An alternative modification to a CMM would be to define a pluggable CMM that provides a standardized extendable control architecture using a plug-in method to provide the implementation of predefined steps. Some of these defined steps might provide, for example, device modeling, gamut mapping, or device channel separation as plug-ins. Default plug-ins can be prescribed for such an implementation, but they can be replaced to meet specific needs. This allows for secret sauce to be implemented in proprietary plug-ins while still providing for some level of baseline openness.

A plug-in can thus be considered an additional form of operational metadata that provides the implementation of transform/control logic not provided directly by the CMM.

In providing plug-ins to a standardized CMM, movement along the CMM/metadata implementation continuum could be considered to be in a different dimension than the static versus dynamic run-time behavior. An additional fixed versus programmable dimension to the CMM/metadata implementation continuum allows comparisons to be made between different levels of plug-in capability of pluggable CMMs. A revised continuum, which replaces that of Figure 6.2, is shown in Figure 6.3.

One serious concern with a pluggable CMM implementation would be that the unambiguous communication of color requires that all CMMs in a complete workflow are configured the same when asked do the exact same task. Do they all have the same plug-ins installed? Is the same essential architecture implemented on different platforms? Are plug-ins implemented (the same) for every platform? Are the plug-ins all configured the same? With pluggable CMM implementations, interoperability is a significant concern.

With this revised version of the CMM/metadata implementation continuum, a static programmable implementation is open for consideration. If the run-time behavior is to be static, then the programmability needs to be fully controlled by the color metadata. Both the color metadata and the CMM need to be extended to provide more operational options, which are controlled by the color metadata. In this case, the operational logic of both the CMM and the color metadata is extended, but the run-time behavior remains static.

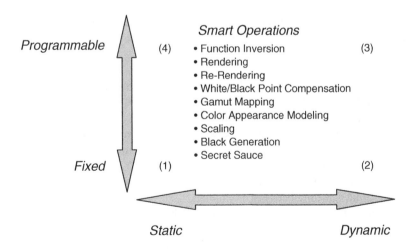

Figure 6.3   Revised CMM/metadata implementation continuum

With a static programmable implementation, greater control and flexibility are possible in an open fashion with the CMM understanding little about what is going on. The additional control is open, as it is added to the color metadata.

A static programmable CMM can be thought of as a general purpose color transform virtual machine (VM) which can easily be ported to different platforms. All that is needed is a specification of the basic building blocks of the VM, and the color metadata can then provide the sequencing to implement various workflows. A static programmable CMM does not necessarily need to understand what the sequence of operations defined in the color metadata is trying to accomplish. Because of this a static programmable CMM can be considered to be a more capable static CMM.

Placing operational sequence control in the color metadata allows for unambiguous *communication* of both *data* and *application* to get *intended* results. For some vendors, the openness may be seen as a weakness – the sequence of operations is openly defined, and any secret sauce is potentially less hidden. However, the increased openness improves the unambiguous communication of color.

In November 2006 the ICC approved the Floating Point Device Encoding Range amendment to the ICC profile specification, which includes of a set of new optional tags that allow for the implementation of a static programmable CMM. See Chapter 31 below for more details.

6.7 Review and Comparisons

For comparison purposes the four corners of the CMM/metadata implementation continuum of Figure 6.3 are now presented with a brief description of general advantages or observations along with disadvantages or concerns.

The points below represent extremes of the implementation continuum, and hybrid approaches will combine features with associated trade-offs. It should of course be recognized here that an advantage to one person might be considered as a disadvantage to another.

1. *Static Fixed*: The operational logic of the transforms involved is placed in fixed sequence in the color metadata. The CMM is responsible for applying the transform steps with limited conversion between transforms.
   Advantages/observations:
   • Most of what needs to be specified is in the metadata specification.
   • CMM specification not as necessary.
   • Easy to make open and cross-platform.
   • Predictability fairly easy to achieve between different implementations.
   • Proprietary know-how is encapsulated/hidden in metadata.
   Disadvantages/concerns:
   • Limited to transform options provided in the specification.
   • Little dynamic run-time behavior is implied.
   • If knowledge of both source and destination is to be used then it is needed at the time when the metadata is created:
     – Knowledge of an intermediary can be used if knowledge of either the source or destination is not known.

  – Use of an intermediary requires that it is well specified and used consistently by
    different implementations
  – Use of an intermediary is not the same as knowing both the source and destination.
  • Limited to features provided in the specification.
2. *Dynamic Fixed*: All operational logic of the transform is placed in the CMM. The color
   metadata only contains characterization/measurement data. Transforms are calculated
   dynamically at run-time.
   Advantages/observations:
   • Proprietary color management requirements may be implemented by proprietary
     CMMs using standard color metadata (Note that, usually, no secret sauce is in the
     color metadata.)
   • The CMM may provide an interface for end-user control of results.
   • Dynamic transform generation allows for transforms to be created based on knowledge of
     data from source and destination as well as image:
     – If knowledge of both source and destination is used then it is not needed until the time
       when the dynamic transformation is generated.
   • Flexibility in metadata/profile connection.
   Disadvantages/concerns:
   • An open solution requires an agreed-upon CMM specification with all operational and
     transform logic being clearly defined and specified:
     – In practice, solutions are usually proprietary for the reasons noted previously, and
       intellectual property issues come to bear.
   • If fixed operational and transform logic is specified, the specification needs to be changed
     to do things differently.
   • Difficult to standardize or to implement the same on many platforms.
   • Predictability between implementations will be difficult due to differences in each
     implementation and how they are configured based upon the opportunity for end-user
     control.
3. *Dynamic Programmable*: The CMM supports a sequence of operations that can be
   customized using a plug-in architecture. The sequence can be scripted or standardized.
   The color metadata contains characterization/measurement data. Operational metadata can
   also potentially be used to determine the sequence of operations and plug-ins to be used.
   Advantages/observations:
   • Greatest flexibility – any color management implementation is possible.
   • Dynamic transform generation allows for custom transforms to be created based on
     knowledge of data from source and destination as well as image.
   • Predetermined transforms can be provided as plug-ins.
   • If knowledge of both source and destination is used then it is not needed until the time
     when dynamic transformation is generated.
   • Depending on implementation, there can be flexibility in metadata/profile connection.
   • Proprietary know-how is placed in plug-ins.
   • Alternative ways of doing things can be encapsulated in plug-ins.
   • Plug-ins can provide interfaces for end-user control of results.
   Disadvantages/concerns:
   • Open solution requires an agreed-upon CMM specification with all transforms clearly
     defined and specified.

- Cross-platform difficulties – plug-ins (in addition to CMM implementations) should be made available for multiple platforms.
- Behavior for default plug-ins needs to be specified and implemented on all platforms for predictability mode to be ensured.
- Workflows crossing multiple systems require that they all support the same plug-in capabilities (where needed) and are configured the same (where needed) based upon the opportunity for end-user control.
- Predictability between implementations will be difficult due to differences in each implementation and how they are configured based upon the potential for end-user control.

4. *Static Programmable*: The CMM acts as a color transform VM. Fixed operations are defined by the metadata specification and implemented in a flexible manner by the CMM. The color metadata provides an arbitrary sequence of operations to be interpreted and executed by the CMM. The CMM does not interpret meaning between operations.

Advantages/observations:

- Most of what needs to be specified is in the specification.
- New workflows and behaviors can be implemented without changes to the CMM.
- Easy to make open and cross-platform.
- Flexibility in metadata/profile connection is possible if the options are in the specification.
- Predictability fairly easy to achieve between different implementations.
- Proprietary know-how is encapsulated/hidden in metadata.

Disadvantages/concerns:

- Repertoire of operations places limitations on programmability.
- Proprietary know-how can become more exposed.
- CMM specification is more of an issue than static fixed.
- Little dynamic run-time behavior is implied.
- If knowledge of both source and destination is used then it is needed at the time when metadata or a profile is created rather than when the metadata/profile is used:
  - Knowledge of an intermediary can be used if knowledge of either the source or destination is not known.
  - Use of an intermediary requires that it is well specified and used consistently by different implementations.
  - Use of an intermediary is not the same as knowing both the source and destination.
- Can the programmable behavior of metadata/profiles invalidate the capabilities of dynamic CMMs that assume fixed transform behavior?

7

ICC Profiles, Color Appearance Modeling, and the Microsoft Windows Color System

A number of color management users have asked about the impact of the Microsoft Windows Color System (WCS) on color management workflows. This chapter attempts to provide some background on WCS and the associated color appearance-based transforms, and compares them to the ICC color management architecture.

WCS uses the CIECAM02 color appearance model, and run-time color rendering, in which the color transforms to be applied to source files are determined after the output devices are known. When using standard ICC profiles in WCS, the software developer has a choice of processing either run-time color rendering with CIECAM02, or color rendering using pre-determined transforms from the ICC profiles.

To understand the differences between WCS and ICC color management, we first need to clarify what is meant by color rendering, gamut mapping, and color appearance models. A gamut mapping operation takes the code values from a source image and converts them to the code values of a reproduction in a way that compensates for differences in the input and output gamut, in terms of both volume and shape. From this perspective, gamut mapping does not include adjusting for preferred colors or adapting colors for different viewing conditions.

On the other hand, a color rendering operation begins with an encoded representation of a scene, and converts that scene representation to a reproduction in a way that includes gamut mapping and image preference adjustments, while also compensating for differences in viewing conditions, dynamic range, and so on. Color re-rendering is similar to color rendering, except that it starts with a source image that is already a reproduction, and produces a new different reproduction, typically for a different kind of display.

A color appearance model uses a parameter set and an algorithm to compute colors encoded in a color appearance color space. The value of a color appearance color space is that it provides a way to represent colors as a human would see them under a particular defined viewing condition. CIECAM02, the color appearance model in WCS, is a well-tested and proven model.

Color Management: Understanding and Using ICC Profiles   Edited by Phil Green
© 2010 John Wiley & Sons, Ltd

Color appearance models such as CIECAM02 are commonly used in the construction of ICC profiles, since they provide a means of predicting the appearance under different viewing conditions and hence predict the colorimetry required to match a source color on a medium which has different viewing conditions.

The idea of run-time color rendering is that the complete color transformation is constructed at run-time, and that the complete transform is specific to the imaging conditions required at that time. ICC member companies have provided various kinds of run-time color rendering solutions to market as far back as the mid- to late 1990s, and enhanced support for run-time color rendering was one of the design goals in the ICC Version 4 revision.

Although ICC profiles can be used to construct run-time color rendering transforms, and some ICC-based applications are available, the dominant modes of ICC operation have used predetermined transforms. This is because quality, predictability, and repeatability have generally been more important to ICC users than run-time output flexibility. Across the markets that the ICC serves, there are business-critical use cases that require the specification of predetermined color behavior: for example, conversion rules to be carried from the design approval point in a workflow to the later implementation stages, and to be archived with digital color files for later matching reproduction. Predetermined transforms encoded in ICC profiles provide this capability.

It is important to keep in mind that a typical color conversion transform, whether it is constructed at run-time or predetermined, will incorporate a number of features. Color appearance models deal with viewing adaptation adjustments between source and destination, but do not address optimization for a variety of output condition particulars, for example, gamut reshaping from monitor to print, printing ink limit, and so on. In many cases the quality of output is determined by the specifics of these optimizations. ICC profiles include pre-optimized transform elements that deal with all aspects of cross-media reproduction. For example, the predetermined perceptual rendering intent transforms in ICC Version 4 profiles are pre-optimized for print production gamut mapping. Version 4 profiles ensure correct interconnection between the predetermined transforms in source and destination profiles through the use of a common, well-defined reference print color gamut.

Because color appearance and color rendering are active areas of research and development, the ICC has chosen not to lock in color management systems to a particular version of a color appearance model. ICC color management, and virtually all color appearance models, are based on CIE colorimetry, which has remained stable since 1931. Basing color source interpretation on CIE colorimetry maintains a consistent color conversion basis for any color rendering algorithm or color appearance model that may be used. For ICC profile users, flexibility in choosing gamut mapping to convert between similar color encodings, color rendering to create an image from a scene, or color re-rendering (e.g., to create an optimized print from a monitor display image) is provided through the rendering intent transforms in ICC profiles and the colorimetric encoding of the PCS. When a run-time color appearance adjustment is required, support is provided by the chromatic adaptation and viewing condition tags in ICC profiles. It is often the case that the predetermined transforms in ICC profiles are the result of extensive expert optimization.

Whenever digital color data is stored or archived, regardless of the processing methods, each image should be stored using a well-defined color encoding and should be tagged with a standard ICC source profile that matches the well-defined color encoding. Saving images and documents tagged with standard ICC profiles in this way will ensure that they can be interpreted correctly on any system, for any use, in the future.

The ICC profile encoding format carries a number of benefits to users, and as a consortium the ICC recognizes that there is a significant installed base of ICC profiles worldwide. A change from the current efficiently encoded, machine-readable format (to a human-readable format such as XML, for example) would place an undue burden on systems and application providers and their customers, while increasing file size and adding no new functionality. The key ICC objective of continuously improving interoperability across open systems motivates against changing the profile format encoding, particularly given that numerous editing and inspection tools that work with the current format are readily available.

Microsoft has stated that WCS will support ICC Version 4 profiles. ICC profiles have become the standard way to interpret the meaning of color values held in digital files and are recognized and processed in hundreds of applications and millions of devices worldwide. The ICC welcomes the new Microsoft WCS support for this community and applauds the work by Microsoft and Canon to advance the state of Windows color management.

In the future, the ICC anticipates broader implementations of dynamic and programmable run-time color rendering in ICC-compliant color management software and devices, along with continuing support of predetermined, fixed transforms using the existing static profile model.

8

Glossary of Terms

This glossary of terms contains definitions of terminology commonly used in color imaging (including digital photography and printing), color reproduction and management, and color and density measurement. Many are taken from the ICC specification, international standards, or CIE publications, and in such cases the specification from which they have been obtained is identified in square brackets. In some cases minor changes have been made, or the notes associated with these definitions removed, either for the purposes of clarity, or to make their use more general. Such changes are indicated by designating the term as "Derived from."

Absolute colorimetric coordinates   tristimulus values, or other colorimetric coordinates derived from tristimulus values, where the numerical values correspond to the magnitude of the physical stimulus. [Derived from ISO 12231]
NOTE 1 When CIE 1931 2° standard observer color matching functions are used, the Y value corresponds to the luminance, not the luminance factor (or some scaled value thereof).
NOTE 2 This should not be confused with the definition of ICC-absolute colorimetry.

Achromatic (perceived) color   color devoid of hue, in the perceptual sense. [CIE publication 17.4, 845-02-26]
NOTE 1 The color names white, gray, and black are commonly used or, for transmitting objects, colorless and neutral.

Adapted white   color stimulus that an observer who is adapted to the viewing environment would judge to be perfectly achromatic and to have a luminance factor of unity; that is, have absolute colorimetric coordinates that an observer would consider to be the perfect white diffuser. [ISO 12231]
NOTE The adapted white may vary within a scene.

Additive RGB color   color formed by mixing light from a set of primary light sources, usually red, green, and blue.

Additive RGB color space   a colorimetric color space having three color primaries (generally red, green, and blue) such that CIE XYZ tristimulus values can be determined from the RGB color space values by forming a weighted combination of the CIE XYZ tristimulus values for the individual color primaries, where the weights are proportional to the radiometrically linear color space values for the corresponding color primaries. [ISO 12231]

NOTE 1 A simple linear 3×3 matrix transformation can be used to transform between CIE XYZ tristimulus values and the radiometrically linear color space values for an additive RGB color space.

NOTE 2 Additive RGB color spaces are defined by specifying the CIE chromaticity values for a set of additive RGB primaries and a color space white point, together with a color component transfer function.

Adopted white   spectral radiance distribution as seen by an image capture or measurement device and converted to color signals that are considered to be perfectly achromatic and to have an observer adaptive luminance factor of unity; that is, color signals that are considered to correspond to a perfect white diffuser. [ISO 12231]

NOTE 1 The adopted white may vary within a scene.

NOTE 2 No assumptions should be made concerning the relation between the adapted or adopted white and measurements of near perfectly reflecting diffusers in a scene, because measurements of such diffusers will depend on the illumination and viewing geometry and on other elements in the scene that may affect perception. It is easy to arrange conditions for which a near perfectly reflecting diffuser will appear to be gray or colored.

Aliasing   output image artifacts that occur in a sampled imaging system for input images having significant energy at frequencies higher than the Nyquist frequency of the system. [ISO 12231]

NOTE These artifacts usually manifest themselves as moiré patterns in repetitive image features or as jagged stair-stepping at edge transitions.

Aligned   a data element is aligned with respect to a data type if the address of the data element is an integral multiple of the number of bytes in the data type. [ICC.1]

Application programming interface (API)   high-level functional description of a software interface. [ISO 12231]

NOTE An API is typically language dependent.

ASCII text string   sequence of bytes, each containing a graphic character from ISO/IEC 646, the last character in the string being a NULL (character 0/0). [ICC.1]

Attribute just noticeable difference (attribute JND)   a measure of the detectability of appearance variations, corresponding to a stimulus difference that would lead to a 75:25 proportion of responses in a paired comparison task in which univariate stimuli pairs were assessed in terms of a single attribute identified in the instructions. [ISO 12231]

NOTE 1 As an example, a paired comparison identifying the sharper of two stimuli that differed only in their generating system modulation transfer function (MTF) would yield results in terms of sharpness attribute JNDs. If the MTF curves differed monotonically and did not cross, the outcome of the paired comparison would depend primarily upon the observers' ability to detect changes in the appearance of the stimuli as a function of MTF variations, with little or no value judgment required of the observers. If a given stimulus difference were genuinely detected by one-half of observers, then on average a 75:25 proportion of responses would result, because those observers detecting the difference would all identify the same sample as being sharper, whereas those not detecting the difference would be forced to guess, and would therefore be equally likely to choose either sample. The relationship between paired comparison proportions and stimulus differences is discussed in greater detail in Annex A of ISO 20462-1.

NOTE 2 If observers are instead asked to choose which of a pair of stimuli is higher in overall image quality, and if the stimuli in aggregate are multivariate, such that the observer must make value judgments of the importance of a number of attributes, rather than focusing on one aspect of image appearance, it is observed experimentally that larger objective stimulus differences (e.g., MTF changes) are required to obtain a 75:25 proportion of responses, which in this case corresponds to a quality JND. In the cases of sharpness and noisiness, approximately twice as large an objective stimulus difference is required to produce one quality JND compared to one attribute JND. Because an attribute change cannot affect quality unless it is detectable, the number of attribute JNDs will always place an upper bound on the number of quality JNDs.

Axis of a (half-tone) screen   one of the two directions in which the half-tone pattern shows the highest number of image elements, such as dots or lines, per length. [ISO 12647-1]

Big-endian   addressing the bytes within a 16-, 32-, or 64-bit value from the most significant to the least significant, as the byte address increases. [ICC.1]

Bit position   bits are numbered such that bit 0 is the least significant bit. [ICC.1]

Bleed   additional printing area outside the nominal printing area necessary for the allowance of mechanical tolerance in the trimming process. [ISO 15930]

NOTE The bleed area includes the area that may be printed but does not include printers' marks of any kind.

Byte   an 8-bit unsigned binary integer. [ICC.1]

Byte offset   number of bytes from the beginning of a field. [ICC.1]

Characterized printing condition   printing condition (offset, newsprinting, publication gravure, flexographic, direct, etc.) for which process control aims are defined and for which the relationship between printing tone values (usually CMYK) and the colorimetry of the printed image is documented. [Derived from ISO 15930]

NOTE 1 The relationship between printing tone values and the colorimetry of the printed image is commonly referred to as characterization.

NOTE 2 It is generally preferred that the process control aims of the printing condition and the associated characterization data be made publicly available via the accredited standards process or industry trade associations.

NOTE 3 Characterization data for many standard printing conditions may be found in the characterization registry on the ICC web site.

Charge-coupled device (CCD)   a type of silicon integrated circuit used to convert light into an electronic signal. [ISO 12231]

Check sum sum of the digits in a file that can be used to check if a file has been transferred properly. [ISO 12640-3]
NOTE Often, only the least significant bits are summed.

Chromatic (perceived) color perceived color possessing hue, in the perceptual sense. [CIE publication 17.4, 845-02-27]

Chromaticity a pair of CIE 1931 x, y values that uniquely describe the hue and saturation (but not the luminance) of a color stimulus.

Chromatic adaptation transform of CIE coordinates to adjust for the appearance change of a stimulus resulting from a change in the chromaticity of the adopted white.

CIELAB color difference; CIE 1976 L^*, a^*, b^* color difference (ΔE_{ab}) difference between two color stimuli defined as the Euclidean distance between the points representing them in L^*, a^*, b^* space. [CIE publication 17.4, 845-03-55]

CIELAB color space; CIE 1976 L^* a^* b^* color space three-dimensional, approximately uniform color space obtained by applying a cube-root transformation to CIE 1931 tristimulus values X, Y, Z, or CIE 1964 tristimulus values X_{10}, Y_{10}, Z_{10}, to obtain L^*, a^*, b^* which are plotted in rectangular coordinates. [Derived from ASTM E284 and CIE publication 17.4, 845-03-56]

CIE XYZ tristimulus values computed using the CIE 1931 Standard Colorimetric Observer.

Color appearance model *(See image appearance model, single stimulus appearance model)*

Color component one of the channels or dimensions in a color data encoding. For example, an RGB encoding has color components red, green, and blue, which are encoded independently of each other.

Color component transfer function single variable, monotonic mathematical function applied individually to one or more color channels of a color space. [ISO 12231]
NOTE 1 Color component transfer functions are frequently used to account for the nonlinear response of a reference device and/or to improve the visual uniformity of a color space.
NOTE 2 Generally, color component transfer functions will be nonlinear functions such as a power-law (i.e., "gamma") function or a logarithmic function. However, in some cases a linear color component transfer function may be used.

Color conversion a transform between color data encodings. In most cases a color conversion results in some property of the source encoding (such as the appearance) being preserved or modified in a systematic way when transformed to the destination encoding.

Color data encoding a quantized digital encoding of a color space.

Color encoding a generic term for a quantized digital encoding of a color space, encompassing both color space encodings and color image encodings. [ISO 22028-1]

Color gamut solid in a color space, consisting of all those colors that are: either present in a specific scene, artwork, photograph, photomechanical, or other reproduction; or capable of being created using a particular output device and/or medium. [ISO 12231]

Color image encoding   digital encoding of the color values for a digital image. [Derived from ISO 12231]

NOTE 1 According to ISO 12231, such encoding must include the specification of a color space encoding (which specifies the encoding method and value range), together with any information necessary to properly interpret the color values such as the image state, the intended image viewing environment, and the reference medium. In some cases the intended image viewing environment will be explicitly defined for the color image encoding. In other cases, the intended image viewing environment may be specified on an image-by-image basis using metadata associated with the digital image. This requirement is essential to properly interpret the color of the data.

NOTE 2 Some color image encodings will indicate particular reference medium character-istics, such as a reflection print with a specified density range. In other cases the reference medium will be not applicable, such as with a scene-referred encoding, or will be specified using image metadata.

NOTE 3 Color image encodings are not limited to pictorial digital images that originate from an original scene, but are also applicable to digital images with content such as text, line art, vector graphics, and other forms of original artwork.

Color management (digital imaging)   communication of the associated data required for unambiguous interpretation of color content data, and application of color data conversions, as required, to produce the intended reproductions. [ICC.1]

NOTE 1 Color content may consist of text, line art, graphics, and pictorial images, in raster or vector form, all of which may be color managed.

NOTE 2 Color management considers the characteristics of input and output devices in determining color data conversions for these devices.

Color matching functions   tristimulus values of monochromatic stimuli of equal radiant power. [CIE Publication 17.4, 845-03-23]

Color rendering   mapping of image data representing the color space coordinates of the elements of a scene or original to output-referred image data representing the color space coordinates of the elements of a reproduction. [Derived from ISO 12231]

NOTE Color rendering generally consists of one or more of the following: compensating for differences in the input and output viewing conditions; tone scale and gamut mapping to map the scene colors onto the dynamic range and color gamut of the reproduction; and applying preference adjustments.

Color re-rendering   mapping of picture-referred image data appropriate for one specified real or virtual imaging medium and viewing conditions to picture-referred image data appropriate for a different real or virtual imaging medium and/or viewing conditions. [See ISO 12231]

NOTE Color re-rendering generally consists of one or more of the following: compensat-ing for differences in the viewing conditions; compensating for differences in the dynamic range and/or color gamut of the imaging media; and applying preference adjustments.

EDITOR'S NOTE From an ICC perspective it may be useful to think of color rendering as a procedure in which there is a change in image state on one side of the PCS (but not both), and color re-rendering as a change in image state on both sides of the PCS. These should be compared to matching in which a colorimetric or appearance match is achieved and there is no change in image state.

Color separation   set of color channels resulting from the conversion of a color file to the printing process colors (usually cyan, magenta, yellow, and black).

Color sequence   order in which the colors are stored in a data file. [ISO 12640-3]

Color space   geometric representation of colors in space, usually of three dimensions. [CIE Publication 17.4, 845-03-25]

Color space encoding   digital encoding of a color space, including the specification of a digital encoding method and a color space value range. [ISO 12231]
NOTE Multiple color space encodings may be defined based on a single color space where the different color space encodings have different digital encoding methods and/or color space value ranges. (For example, 8-bit sRGB and 10-bit e-sRGB are different color space encodings based on a particular RGB color space.)

Color space white point   color stimulus to which color space values are normalized. [ISO 12231]
NOTE The color space white point may or may not correspond to the assumed adapted white point and/or the reference medium white point for a color image encoding.

Color value   numeric values associated with each of the pixels of an image, or each point of a color space. [Derived from ISO 12640-3]

Colorant   substance that modifies the color of a substrate, usually a dye or pigment.

Colorimeter   instrument for measuring colorimetric quantities, such as the tristimulus values of a color stimulus. [CIE publication 17.4, 845-05-18]. *(See spectrocolorimeter and tristimulus colorimeter)*

Colorimetric color space   color space having an exact and simple relationship to CIE colorimetric values. [ISO 12231]
NOTE Colorimetric color spaces include those defined by CIE (e.g., CIE XYZ, CIELAB, CIELUV, etc.), as well as color spaces that are simple transformations of those color spaces (e.g., additive RGB color spaces).

Control patch   area produced for control or measurement purposes. [ISO 12647-1]

Control strip   one-dimensional array of control patches. [ISO 12647-1]

Data range   range of values for a given variable in between a minimum and maximum value. [Derived from ISO 12640-3]

Depth of field   difference between the maximum and minimum distances from a camera lens's front nodal point to objects in a scene that can be captured in acceptably sharp focus. [ISO 12231]

Deviation tolerance   permissible difference between the OK print from a production run and the reference value. [ISO 12647-1]

Device   a system capable of recording or producing color stimuli.

Device characterization   the process of defining the relationship between device values and tristimulus values, or their derivatives.

Device-dependent color space   color space defined by the characteristics of a real or idealized imaging device. [ISO 12231]
NOTE Device-dependent color spaces having a simple functional relationship to CIE colorimetry can also be categorized as colorimetric color spaces. For example, additive RGB color spaces corresponding to real or idealized CRT displays can be treated as colorimetric color spaces.

Diffuse reflection   diffusion by reflection in which, on the macroscopic scale, there is no regular reflection. [CIE Publication 17.4, 845-04-47]

Digital imaging system   system that records and/or produces images using digital data. [ISO 12231]

Digital output level   numerical value assigned to a particular output level, also known as the digital code value. [ISO 12231]

Doubling/slur patch   control patch for the assessment of the true rolling condition. [ISO 13655]

Dynamic range   the ratio of the minimum to the maximum intensities in a system.

Electro-optical conversion function (EOCF)   relationship between the digital code values provided to an output device and the equivalent neutral densities produced by the device. [ISO 12231]

Engraving pitch (*P*)   cell spacing on a gravure cylinder, evaluated from the following formula:

$$P = \sqrt{a \times b}$$

where:

a is the distance between the same points on two adjacent cells in the printing direction;
b is the distance between adjacent circumferential tracks of the engraving stylus.
[ISO 12647-4]

EPS   Encapsulated PostScript as defined by Adobe Technical Note #5002. [ISO 15930]

Equivalent neutral density (END)   visual density or effective visual density of an analysis primary or rendering colorant, when it is combined with the amounts of the other system primaries or colorants required to produce a visual neutral. [ISO 12231]

Exchangeable image file format (Exif)   the standard for the digital still camera image file format of the Japan Electronic Industry Development Association (JEIDA). [ISO 12231]
NOTE The JPEG version of Exif provides a compressed file format for digital cameras in which the images are compressed using the baseline JPEG standard described in ISO/IEC 10918-1, and metadata and thumbnail images are stored using TIFF tags within an application segment at the beginning of the file.

Film rendering transform   mapping of image data representing measurements of a photographic negative to output-referred image data representing the color space coordinates of the elements of a reproduction. [ISO 12231]

Film unrendering transform   mapping of image data representing measurements of a photographic negative to scene-referred image data representing estimates of the color space coordinates of the elements of the original scene. [ISO 12231]

Fixed point   method of encoding a real number into binary by putting an implied binary point at a fixed bit position. [ICC.1]

NOTE Many of the tag types defined in this international standard contain fixed point numbers. Several references can be found (MetaFonts etc.) illustrating the preferability of fixed point representation to pure floating point representation in very structured circumstances.

Flare   light falling on an image, in an imaging system, which does not emanate from the subject point. [ISO 12231]

NOTE Veiling glare is also sometimes referred to as flare.

Floating point   representation of a real number by a string of digits, in which the radix (decimal point) can be placed at any location within the string. Binary floating point encodings are defined in IEEE 754 with a sign bit, an exponent, exponent offset (or bias), and significant (or mantissa).

Gamma correction   signal processing operation that changes the relative signal levels in order to adjust the image tone reproduction. [Derived from ISO 12231]

NOTE 1 Gamma correction is normally used in the context of displays, and is performed in part to correct for the nonlinear light output versus signal input characteristic of the display. The relationship between the light input level and the output signal level, called the OECF, provides the gamma correction curve shape for an image capture device. The relationship between the output signal level and equivalent neutral density produced, called the EOCF, provides the gamma correction curve shape for an output device.

NOTE 2 The gamma correction is usually an algorithm, look-up table, or circuit which operates separately on each color component of an image.

Gamut mapping   mapping of the color space coordinates of the elements of a source image to the color space coordinates of the elements of a reproduction to compensate for differences in the source and output medium color gamut capability. [Derived from ISO 12231]

NOTE The term "gamut mapping" is somewhat more restrictive than the term "color rendering" because gamut mapping is performed on colorimetry which has already been adjusted to compensate for viewing condition differences and viewer preferences, although these processing operations (gamut mapping and color rendering) are frequently combined in reproduction and preferred-reproduction models.

Global color change   change to the colors in an image (often selectively by color region) applied consistently to all parts of the image. [ISO 12640-3]

NOTE This is in contrast to a local color change where selected spatial areas of an image are changed separately from the rest of the image area.

Gloss   mode of appearance by which reflected highlights of objects are perceived as super-imposed on the surface due to the directionally selective properties of that surface. [CIE Publication 17.4, 845-04-73]

Gray balance   set of tone values for cyan, magenta, and yellow on the color separation films, or in a data file, that appears as an achromatic color under specified viewing conditions when printed under specified printing conditions. [Derived from ISO 12647-1]
NOTE Two practical procedures are often used to define such a gray: "a color having the same CIELAB a^* and b^* values as the print substrate" and "a color that has the same CIELAB a^* and b^* values as a half-tone tint of similar L^* value printed with black ink." Both usually produce different values, and both may differ somewhat from a perceived gray seen in isolation.

Gray component replacement (GCR)   replacement of some, or all, of cyan, magenta, and yellow printing inks by black ink, such that the color is maintained. *(See under color removal (UCR))*

Half-tone   image composed of dots which can vary in screen ruling (number per centimeter), size, shape, or density, thereby producing tonal gradations. [ISO 12637-1]

Hard copy   representation of an image on a substrate which is self-sustaining and reasonably permanent. [ISO 12231]

Hexadecimal   number system used to represent the value of a 4-bit binary word. [ICC.1]
NOTE The notation used to represent hexadecimal numbers in the ICC specification is xxh.

ICC color management   method for the controlled conversion of color data from one color encoding to another, by means of ICC profiles. [Derived from ISO 12231 and ISO 12647-1]

ICC (International Color Consortium)   industry body responsible for the ICC profile specification and color management architecture. [ISO 12647-1]

ICC profile   set of transforms from one color encoding to another, for example, from device color coordinates to profile connection space, prepared in accordance with ICC.1. [Derived from ISO 12231 and ISO 12647-1]

Illuminant   spectral power distribution defined over the wavelength range that influences object color perception.

Image appearance model   mathematical model that uses information about viewing conditions and other colors in an image to estimate the subjective appearance of any element of an image from colorimetric measurements of the image. [Derived from ISO 12231]
NOTE 1 An image appearance model can describe the elements of a scene, an original, or a reproduction, but it does not consider the characteristics of any potential output medium for subsequent reproduction. (EDITOR'S NOTE The characteristics of the output medium may be relevant to an appearance model if the medium itself influences the visual state – such as chromatic adaptation.)
NOTE 2 There is no general consensus on the appropriate form for an image appearance model. There is an expectation that reproducing the colorimetric coordinates of every color in an image will result in a reproduction of the appearance of the entire image, as long as the

viewing condition and size of the image remain the same, and it is possible to reproduce the colorimetric coordinates of every color in the image.

NOTE 3 An image appearance model followed by its inverse is appropriate for use as a reproduction model provided the reproduction medium does not impose any limitations on the colors to be reproduced in the image.

Image capture device device for converting a scene, or a fixed image such as a print, film, or transparency, to digital image data. [ISO 12231]

Image data format structure and content which specifies how the data is logically organized on a given storage medium. [ISO 12231]

Image noise unwanted variations in an image. [ISO 12231]

Image output device device that can render a digital image to hard copy or soft copy media. [ISO 12231]

Image state attribute of a color image encoding indicating the type of image data to which the encoded image color values are referred. [ISO 12231]

NOTE The image states defined in this glossary are scene-referred, original-referred, and output-referred states.

Improved newsprint paper with, compared to ordinary newsprint, a higher smoothness, a higher brightness, and a filler content up to 20%. [ISO 12647-4]

Incident flux flux incident on the sampling aperture defining the specimen area on which the measurement is made. [ISO 5-1]

Ink trap for an overprint, a relative measure for the average amount of colorant per unit area of the second-down colorant layer that is deposited onto the first-down colorant layer. To be expressed as a percentage. [ISO 13655]

NOTE 1 Not to be confused with trap employed in color separation to attenuate mis-register effects.

NOTE 2 Apparent ink trap is measured optically; gravimetric ink trap by weight.

Interpolation calculating the value of a function which lies between already-known values of the function.

Luminance the luminous intensity of light emitted by a surface. It is expressed as the luminous flux emitted divided by the area of the surface, usually in candelas per square meter.

Luminance factor ratio of the luminance of the surface element in the given direction to that of a perfect reflecting or transmitting diffuser identically illuminated. [CIE Publication 17.4, 845-04-69]

Magnitude estimation method a psychophysical method involving the assignment of a numerical value to each test stimulus that is proportional to the attribute under test, such as an attribute of image quality. [Derived from ISO 12231]

NOTE 1 Typically, a reference stimulus with an assigned numerical value is present to anchor the rating scale.

NOTE 2 The numerical scale resulting from a magnitude estimation experiment is usually assumed to constitute a ratio scale, which, ideally, is a scale in which a constant percentage change in value corresponds to one JND. In practice, modest deviations from this behavior occur, complicating the transformation of the rating scale into units of JNDs without inclusion of unidentified reference stimuli (having known quality) among the test stimuli.

Medium   the combination of device, colorants, and substrate which form a color stimulus produced by a color output system.

Medium black point   lowest luminance neutral that can be produced by an imaging medium in normal use, measured using the specified measurement geometry. [ISO 12231]
NOTE It is generally desirable to specify a medium black point that has the same chromaticity as the medium white point.

Medium white point   highest luminance neutral that can be produced by an imaging medium in normal use, measured using the specified measurement geometry. [ISO 12231]

Metadata   data associated with a digital image aside from the pixel values that comprise the digital image. [ISO 12231]
NOTE Metadata is typically stored as tags in the digital image file

Mid-tone spread *(S)*   quantity defined by the following equation:

$$S = \max\{(Ac - Ac0), (Am - Am0), (Ay - Ay0)\} - \min\{(Ac - Ac0), (Am - Am0), (Ay - Ay0)\}$$

where:

Ac is the measured tone value of the cyan process color image;
Ac0 is the specified tone value of the cyan process color image;
Am is the measured tone value of the magenta process color image;
Am0 is the specified tone value of the magenta process color image;
Ay is the measured tone value of the yellow process color image;
Ay0 is the specified tone value of the yellow process color image.

EXAMPLE The calculation of the mid-tone spread:

measured values$(c, m, y) = (22, 17, 20)$; specified values$(c, m, y) = (20, 20, 18)$

$\max\{(22 - 20), (17 - 20), (20 - 18)\} = 2; \min\{(22 - 20), (17 - 20), (20 - 18)\} = -3$

$S = \{\max - \min\} = 5.$

[ISO 12647-1]

Moiré pattern   unwanted periodic structure produced by interference between two or more two-dimensional periodic structures. [ISO 12647-1]

Most significant nibble (MSN)   most significant 4 bits of the most significant byte. [ISO 12231]

Non-periodic (half-tone) screen   a half-tone screen without a regular half-tone dot pattern. [Derived from ISO 12647-1]
NOTE Also known as a stochastic or frequency-modulated (half-tone) screen.

NULL   character coded in position 0/0 of ISO/IEC 646. [ICC.1]

Nyquist limit   spatial frequency equal to 0.5 times the inverse of the sampling period. [ISO 12231]
NOTE Energy at input spatial frequencies above the Nyquist limit will alias to a spatial frequency below the Nyquist limit in the output image. The Nyquist limit may be different in the two orthogonal directions.

Off-press proof print   a print produced by a method other than press printing whose purpose is to show the results of the color separation process in a way that closely simulates the results on a production press. [ISO 12647-1]
NOTE Also known as artificial or prepress proof.

OK print; OK sheet   during production printing the production print singled out as reference for the remaining production run. [ISO 12647-1]

On-press proof print   print produced by press printing (production or proof press) whose purpose is to show the results of the color separation process in a way that closely simulates the results on a production press. [ISO 12647-1]

Opto-electronic conversion function (OECF)   relationship between the log of the input levels and the corresponding digital output levels for an opto-electronic digital image capture system. [ISO 12231]
NOTE If the input log exposure points are very finely spaced and the output noise is small compared to the quantization interval, the OECF may have a step-like character. Such behavior is an artifact of the quantization process and should be removed by using an appropriate smoothing algorithm or by fitting a smooth curve to the data.

Orientation   specifies the origin and direction of the first line of data, with respect to the image content as viewed by the end user. [Derived from ISO 12640-3]

Original-referred image data   image data which represents the color space coordinates of the elements of a two-dimensional hard copy or soft copy image, typically produced by scanning artwork, photographic transparencies or prints, or photomechanical or other reproductions. [ISO 12231]
NOTE 1 Original-referred image data is related to the color space coordinates of the original as measured according to ISO 13655, and does not include any additional veiling glare or other flare.
NOTE 2 The characteristic of original-referred image data that most generally distinguishes it from scene-referred image data is that it is referred to a two-dimensional surface, and the illumination incident on the two-dimensional surface is assumed to be uniform (or the image data corrected for any non-uniformity in the illumination).
NOTE 3 There are classes of originals that produce original-referred image data with different characteristics. Examples include various types of artwork, photographic prints, photographic transparencies, emissive displays, and so on. When selecting a color re-rendering algorithm, it is usually necessary to know the class of the original in order to determine the appropriate color re-rendering to be applied. For example, a colorimetric intent is generally applied to artwork, while different perceptual algorithms are applied to produce photographic prints from transparencies, or newsprint reproductions from

photographic prints. In some cases the assumed viewing conditions are also different between the original classes, such as between photographic prints and transparencies, and will usually be considered in well-designed systems.

NOTE 4 In a few cases, it may be desirable to introduce slight colorimetric errors in the production of original-referred image data, for example, to make the gamut of the original more closely fit the color space, or because of the way the image data was captured (such as a Status A densitometry-based scanner).

Output-referred image data   the image data which represents the color space coordinates of the elements of an image that has undergone color rendering appropriate for a specified real or virtual output device and viewing conditions. [ISO 12231]

NOTE 1 The output-referred image data is referred to the specified output device and viewing conditions. A single scene can be color rendered to a variety of output-referred representations depending on the anticipated output viewing conditions, media limitations, and/or artistic intents.

NOTE 2 Output-referred image data may become the starting point for a subsequent reproduction process. For example, sRGB output-referred image data is frequently considered to be the starting point for the color re-rendering performed by a printer designed to receive sRGB image data.

Overprint   condition where two or more layers of colorant, usually ink, are printed on top of each other. [ISO 13655]

Paired comparison method   a psychophysical method involving the choice of which of two simultaneously presented stimuli exhibits greater or lesser image quality or an attribute thereof, in accordance with a set of instructions given to the observer. [ISO 12231]

NOTE Two limitations of the paired comparison method are as follows. (1) If all possible stimulus comparisons are done, as is usually the case, a large number of assessments are required for even modest numbers of experimental stimulus levels (if N levels are to be studied, $N(N-1)/2$ paired comparisons are needed). (2) If a stimulus difference exceeds approximately 1.5 JNDs, the magnitude of the stimulus difference cannot be directly estimated reliably because the response saturates as the proportions approach unanimity. However, if a series of stimuli having no large gaps are assessed, the differences between more widely separated stimuli may be deduced indirectly by summing smaller, reliably determined (unsaturated) stimulus differences. The various methods for the transformation of paired comparison data to an interval scale (a scale linearly related to JNDs) perform statistically optimized procedures for inferring the stimulus differences using all the available data in a weighted estimate.

PDF (Portable Document Format)   file format defined in the Adobe Portable Document Format. [ISO 15930]

Perceptual Reference Medium (PRM)   a virtual medium encoding with a specified dynamic range and color gamut.

Perceptual Reference Medium Gamut (PRMG)   a PRM which encompasses the majority of surface colors that may be encountered in reflection print color reproduction. The coordinates of the PRMG are defined in Section 2.2 of ISO 12640-3.

Perfect diffusing reflector, perfect diffuser   a surface that reflects incident radiance isotropically. The brightness of a perfect diffuser appears the same at all angles.

Photography   acquisition, processing, or reproduction of optically formed images using chemical or electronic technologies. [ISO 12231]

Pixel   smallest discrete picture element in a digital image file. [ISO 12640-3]

Pixel interleaving   image data organized such that the color values for one pixel are followed by the same sequence of color values for the next pixel. [Derived from ISO 12640-3]
NOTE Other forms of color data interleaving are line and plane.

Preferred-reproduction model   mathematical model that produces transformations which are applied to scene- or original-referred image data to produce image data describing a pleasing reproduction. [ISO 122321]
NOTE Preferred-reproduction models are different from reproduction models in that the pleasing reproduction need not be an attempt to reproduce the appearance of the original. In fact, what is considered pleasing may depend on viewer preferences. The transformations produced by a preferred-reproduction model are generally dependent on the characteristics of the scene or original and the output medium.

Preview   a simulation of the appearance of a stimulus, on a different device or medium to that used to generate the target output

Primary colors   colors of individual prints produced from yellow, magenta, and cyan inks. [Derived from ISO 2846-1]
NOTE If the prints are produced as specified in the appropriate part of ISO 2846, and conform to the colorimetric characteristics specified in that specification, the colors are standard primary colors.

Principal axis (of a half-tone screen)   axis of a screen that coincides with the direction of the longest diameter of an oblong-shaped (e.g., elliptical or diamond-shaped) half-tone dot. [ISO 12647-1]
NOTE Circular and square-shaped half-tone dots do not have a principal axis.

Print substrate   material bearing the printed image. [ISO 12647-1]

Printing condition   set of printing details which fully describe the conditions associated with a specific printed output, usually associated with characterization data measured from an ISO 12642 or similar target. [ISO 12647-1]
NOTE Such parameters usually include (as a minimum) printing process, print substrate type, printing ink, screen type and screen frequency, manner used to produce the printing forme, and surface finish.

Printing forme   tool whose surface is prepared such that some parts transfer printing ink whereas other parts do not. [ISO 12647-1]

Printing tone value   data value corresponding to the relative area of a printing surface that is intended to transfer ink to the substrate being printed. [ISO 15930]

Process color solid   printed area of a process color that corresponds to 100% tone value, or the maximum cell volume identified for the combination of gravure engraving parameters. [Derived from ISO 12647-4]

Process colors (for four-color printing)   yellow, magenta, cyan, and black. [ISO 12647-1]

Profile connection space (ICC PCS)   abstract color space defined by the International Color Consortium providing a standard connection point for combining ICC source and destination profiles. [Derived from ICC.1 and ISO 12231]
NOTE Unlike earlier versions of the specification, there are two variations of the PCS defined in the current ICC specification. One is an original-referred variation for colorimetric intent profiles, the other is a standard output-referred variation for perceptual intent profiles.

Profile creator   a program or utility that takes as input numeric values which sample a color data encoding, calculates the data to be stored in a profile, and then writes a profile according to the ICC specification.

Psychophysical experimental methods   experimental technique for subjective evaluation of image quality or attributes thereof, from which stimulus differences in units of JNDs may be estimated. [ISO 12231]

Quality just noticeable difference (quality JND)   a measure of the significance or importance of quality variations, corresponding to a stimulus difference that would lead to a 75:25 proportion of responses in a paired comparison task in which multivariate stimuli pairs were assessed in terms of overall image quality. [ISO 12231]
NOTE The attribute JND is a measure of detectability of appearance changes, whereas the quality JND is a measure of significance or importance of stimulus differences in terms of their impact on quality. An attribute JND is a useful unit for predicting how observers would react to an advertisement showing images carefully matched in all respects but one, and drawing the attention of the observer to the attribute varying. In contrast, a quality JND is useful for predicting how observers would perceive overall quality as a function of one or more stimulus variations, and so is a more useful quantity in optimizing imaging system design, where different attributes must be balanced against one another. The overall quality of an image may be predicted from knowledge of the impact of each attribute in isolation, expressed in terms of quality JNDs, whereas the same is not true of attribute JNDs. Therefore, it is often highly desirable to obtain results expressed in quality JNDs, even if the stimuli being assessed are univariate in nature. This can be accomplished if test stimuli are rated against a series of appropriately calibrated reference stimuli, as in the quality ruler method.

Quality ruler method   a psychophysical method described in ISO 20462-3, which involves quality or attribute assessment of a test stimulus against a series of ordered, univariate reference stimuli that differ by known numbers of JNDs. [ISO 12231]

Quantization   the representation of a continuous quantity as a finite set of discrete values.

Rank ordering method   a psychophysical method involving the arrangement by an observer of a series of stimuli in order of increasing or decreasing image quality or an attribute thereof, in accordance with the set of instructions provided. [ISO 12231]

Raw digital still camera (DSC) image data   the image data produced by, or internal to, a DSC that has not been processed, except for A/D conversion and the following optional steps: linearization, dark current/frame subtraction, shading and sensitivity (flat field) correction, flare removal, white balancing (e.g., so the adopted white produces equal RGB values or no chrominance), missing color pixel reconstruction (without color transformations). [ISO 12231]

Reader   application, system, or subsystem that accepts a file as its input and performs a level of processing on the file that, at a minimum, accepts or rejects the file based on predetermined criteria and, if accepted, passes the file to the next stage of processing. [ISO 12639]

Reference stimulus   an image provided to the observer for the purpose of anchoring or calibrating the perceptual assessments of test stimuli in such a manner that the given ratings may be converted to JND units. [ISO 12231]

Reflectance factor   ratio of the measured reflected flux from the specimen to the measured reflected flux from a perfect-reflecting and perfect-diffusing material located in place of the specimen. [ISO 5-4]

Reflection densitometer   instrument which measures reflectance factor density. [ISO 12647-1]

Reflection density (or reflectance factor (optical) density)   logarithm to base 10 of the reciprocal of the reflectance factor. [ISO 5-4 and CIE Publication 17.4, 845-04-67]

Reflectometer   photometer for measuring quantities pertaining to reflection. [CIE Publication 17.4, 845-05-26]

Regular reflection (or specular reflection)   reflection in accordance with the laws of geometrical optics, without diffusion. [CIE Publication 17.4, 845-04-45]

Relative density   the density from which the density of a reference such as the film base, or the unprinted print substrate, has been subtracted. [ISO 12647-1]

Rendering intent   style of mapping color values from one image description to another. [Derived from ICC.1]
NOTE ICC defines four rendering intents (ICC-absolute colorimetric, relative colorimetric, perceptual, and saturation).

Reproduction model   mathematical model that produces transformations which are applied to scene- or original-referred image data to produce image data describing a reproduction which is as close as possible to being an appearance match to the original. [ISO 12231]
NOTE Transformations produced by reproduction models will generally depend on the luminance ratio and color gamut of the scene or original and the output medium.

Re-purposing   processing encoded image data to produce a different representation of the image, usually to produce a different output-referred representation and/or to produce a different artistic intent.

Resolution   measure of the ability of a digital image capture system, or a component of a digital image capture system, to depict spatial picture detail. [ISO 12231]

NOTE There are numerous resolution measurement metrics, including resolving power, visual resolution, and limiting resolution.

Re-targeting processing encoded image data to produce a similar representation of the image, usually to produce an output-referred representation with similar appearance on a medium with a different color gamut.

Sample spacing physical distance between sampling points or sampling lines. [ISO 12231]
NOTE The sample spacing may be different in the two orthogonal sampling directions.

Sampled imaging system imaging system or device which generates an image signal by sampling an image at an array of discrete points, or along a set of discrete lines, rather than a continuum of points. [ISO 12231]
NOTE The sampling at each point is done using a finite size sampling aperture or area.

Sampling aperture area of the sample that contributes to the measurement. [ISO 13655]
NOTE This is not necessarily the same as the illumination aperture, which is the area of the sample illuminated by the instrument or the mechanical aperture created by an opaque mask used to position the densitometer on the specimen. ISO 5–4 makes very specific requirements on the relationship between each of these.

Sampling aperture size dimensions of the surface area of the sample that contributes to the measurement of the reflectance or transmittance factor (or reflection or transmission density), governed by the design of the instrument. [Derived from ISO 12647-1]

Sampling frequency reciprocal of the sample spacing. [ISO 12231]
NOTE The sampling frequency is expressed in samples per unit distance.

Scanner electronic device that converts a fixed image, such as a print or film transparency, into an electronic signal. [ISO 12231]

Scene spectral radiances of a view of the natural world as measured from a specified vantage point in space and at a specified time. [ISO 12231]
NOTE A scene may represent an actual view of the natural world or a computer-generated simulation of such a view.

Scene luminance ratio ratio of the highest (highlight) luminance value to the lowest (shadow) luminance value in a scene. [ISO 12231]

Scene-referred image data image data which represents estimates of the color space coordinates of the elements of a scene. [ISO 12231]
NOTE 1 Scene-referred image data can be determined from raw DSC image data before color rendering is performed. Generally, DSCs do not write scene-referred image data in image files, but some may do so in a special mode intended for this purpose. Typically, DSCs write standard output-referred image data where color rendering has already been performed.
NOTE 2 Scene-referred image data typically represents relative scene colorimetry estimates. Absolute scene colorimetry estimates may be calculated using a scaling factor. The scaling factor can be derived from additional information such as the image OECF, FNumber or ApertureValue, and ExposureTime or ShutterSpeedValue tags.
NOTE 3 Scene-referred image data may contain inaccuracies due to the dynamic range limitations of the capture device, noise from various sources, quantization, optical blurring

and flare that are not corrected for, and color analysis errors due to capture device metamerism. In some cases, these sources of inaccuracy can be significant. ISO 17321-1 specifies a DSC/SMI (Sensitivity Metamerism Index), which can be used to estimate the amount of inaccuracy resulting from capture device metamerism.

NOTE 4 The transformation from raw DSC image data to scene-referred image data depends on the relative adopted whites selected for the scene and the color space used to encode the image data. If the chosen scene adopted white is inappropriate, additional errors will be introduced into the scene-referred image data. These errors may be correctable if the transformation used to produce the scene-referred image data is known, and the color encoding used for the incorrect scene-referred image data has adequate precision and dynamic range.

NOTE 5 Standard methods for the calculation of scene-referred image data from raw DSC image data will be specified in ISO 17321-2.

NOTE 6 The scene may correspond to an actual view of the natural world, or a computer-generated simulation of such a view. It may also correspond to a modified scene determined by applying modifications to an original scene to produce some different desired scene. Any such scene modifications should leave the image in a scene-referred image state, and should be done in the context of an expected color rendering transform.

Screen angle   with oblong-shaped half-tone dots, the angle which the principal axis of the screen makes with the reference direction; with circular and square dot shapes, the smallest angle which an axis of the screen makes with the reference direction. [ISO 12647-1]

Screen frequency, screen ruling   number of image elements, such as dots or lines, per unit of length in the direction which produces the highest value. [ISO 12647-1]

Screen width   reciprocal of screen ruling. [ISO 12647-1]

Secondary (ink) colors   colors obtained by overprinting pairs of the three chromatic inks. [Derived from ISO 2846-1]

(ICC-registered) signature   alphanumeric 4-byte value, registered with the ICC. [Derived from ICC.1]

Single stimulus appearance model   mathematical model which uses information about viewing conditions to estimate the subjective appearance of a colored patch from colorimetric measurements of that patch and its surround. [Derived from ISO 12231]

NOTE A single stimulus appearance model cannot be expected to deal completely with the effect of changing viewing conditions in an image, because the combined effect of macroscopic viewing conditions and other colors in the image could result in the appearance of any color in the image changing in a way that is not predictable by the single stimulus model, since it is not keeping track of the other colors.

Soft copy   representation of an image produced using a device capable of directly representing different digital images in succession and in a non-permanent form, the most common example being a display. [Derived from ISO 12231]

Solid   image of uniform coloration intensity with no half-tone structure. [ISO 13656]

Spectral product   product of the spectral power of the incident flux and the spectral response of the receiver, wavelength by wavelength. [ISO 13655]

Spectral response (of the receiver)   product of the spectral sensitivity of the photodetector and the transmittance of the optical elements associated with it. [ISO 13655]

Spectrally non-selective (or spectrally neutral)   exhibiting reflective or transmissive characteristics which are constant over the wavelength range of interest. [ISO 12231]

Spectrocolorimeter   colorimeter which achieves the measurement values by calculation from the spectral data. [Derived from ISO 12647-1]

Specular   reflection of light at an angle that is equal (with respect to the surface normal) to the incident angle. Smooth, glossy, and mirror-like surfaces have a high ratio of specular to diffuse reflection.

Spot color   single colorant, identified by name, whose printing tone values are specified independently from the color values specified in a color coordinate system. [ICC.1 and ISO 15930]

Standard (process) ink   ink, intended for four-color printing, which when printed on the reference substrate and within the applicable range of ink film thicknesses complies with the colorimetric and transparency specifications of the relevant part of ISO 2846. [Derived from ISO 2846-1]

Standard (process) ink set   complete set of standard (process) inks comprising yellow, magenta, cyan, and black. [Derived from ISO 2846-1]

Standard Quality Scale (SQS)   a fixed numerical scale of quality defined in ISO 20462-3 and having the following properties: (1) the numerical scale is anchored against physical standards; (2) a one unit increase in scale value corresponds to an improvement of one JND of quality; and (3) a value of zero corresponds to an image having so little information content that the nature of the subject of the image is difficult to identify. [ISO 12231]

Stimulus   an image presented or provided to the observer either for the purpose of anchoring a perceptual assessment (a reference stimulus) or for the purpose of subjective evaluation (a test stimulus). [ISO 12231]

Subtractive color space   color space obtained by combining colorants which absorb some of the light reflected or transmitted by a substrate. Typical colorants are cyan, magenta, and yellow, with the addition of black in many printing applications.

Surface finishing   process by which a print is either covered by varnish (lacquer) or laminated with a transparent polymeric film. [ISO 12647-1]

Surround   in color appearance, the surround is the field beyond the immediate background to a stimulus. In a color appearance model, the surround is defined as the ratio of the luminance of a white under the surround conditions to the luminance of the media white.
NOTE In some publications the term "surround" denotes the field adjacent to the stimulus.

TIFF   a tagged image file format as defined by revision 6.0 of the TIFF specification. [ISO 15930]

Tonal compression transform which compensates for differences in dynamic range, so that the entire dynamic range of the input medium can be mapped to the dynamic range of the output medium.

Tone reproduction relationship of one of the luminance, luminance factor, L^*, decadic logarithm of luminance, or density in a scene or original to one of the luminance, luminance factor, L^*, decadic logarithm of luminance, or density in a reproduction. [ISO 12231]

NOTE 1 It is not necessary for corresponding quantities to be plotted together, although generally linear quantities are not plotted with respect to logarithmic quantities, and vice versa.

NOTE 2 The term "tone reproduction," also called "system tone reproduction," should only be applied to those processes that both start and end with a visible image. A visible image may be a scene, hard copy, or soft copy. The term "tone reproduction" should not be used with respect to the characteristics of either an input or an output device taken by itself; such devices are system components but they are not systems. So, for example, an OECF of an input device or the tone scale of an output device (such as a printer characteristic curve) do not exemplify tone reproduction.

Tone value (in a data file) (A) proportional printing value encoded in a data file and interpreted as defined in the file format specification.

NOTE Most files store this data as 8-bit integer values, that is, 0–255. The tone value of a pixel is typically computed from the equation

$$A\% = 100 \times \left(\frac{V_p - V_0}{V_{100} - V_0}\right)$$

where:

V_p is the integer value of the pixel;
V_0 is the integer value corresponding to a tone value of 0%;
V_{100} is the integer value corresponding to a tone value of 100%.

[ISO 12647-1]

Tone value; dot area (on a half-tone film of negative polarity) (A) percentage calculated from

$$A\% = 100 \times \left(1 - \frac{1 - 10^{-(D_t - D_0)}}{1 - 10^{-(D_s - D_0)}}\right)$$

where:

D_0 is the transmittance density of the clear half-tone film;
D_s is the transmittance density of the solid;
D_t is the transmittance density of the half-tone.

[ISO 12647-1]
NOTE Formerly known as the film printing dot area.

Tone value; dot area (on a half-tone film of positive polarity) (*A*)   percentage calculated from

$$A\% = 100 \times \left(\frac{1 - 10^{-(D_t - D_0)}}{1 - 10^{-(D_s - D_0)}} \right)$$

where:

D_0 is the transmittance density of the clear half-tone film;
D_s is the transmittance density of the solid;
D_t is the transmittance density of the half-tone.

[Derived from ISO 12647-1]
NOTE 1 Formerly known as the film printing dot area.
NOTE 2 The above equation is often known as the Murray–Davies equation.

Tone value; dot area (on a print) (*A*)   percentage of the surface which appears to be covered by colorant of a single color (if light scattering in the print substrate and other optical phenomena are ignored), calculated from

$$A\% = 100 \times \left(\frac{1 - 10^{-(D_t - D_0)}}{1 - 10^{-(D_s - D_0)}} \right)$$

where:

D_0 is the reflectance factor density of the unprinted print substrate, or the non-printing parts of the printing forme;
D_s is the reflectance factor density of the solid;
D_t is the reflectance factor density of the half-tone.

[Derived from ISO 12647-1]
NOTE 1 Formerly also known as apparent, equivalent, or total dot area.
NOTE 2 The synonym "dot area" may be applied only to half-tones produced by dot patterns.
NOTE 3 This definition may be used to provide an approximation of the tone value on certain printing formes.
NOTE 4 The above equation is often known as the Murray–Davies equation.

Tone value increase; dot gain   difference between the tone value on the print and the tone value on the half-tone film or in the digital data file. [ISO 12647-1]
NOTE The synonym "dot gain" may be applied only to half-tones produced by dot patterns.
EXAMPLE 1 The tone value of the control strip patch on the print is 55%; that on the film is 40%. The tone value increase is 15%.
EXAMPLE 2 The tone value of a flat tint produced by an application program is set to be 75%; the corresponding tint on the print is measured at 92%. The tone value increase is 17%.

Tone value sum   sum of the tone values, at a given image spot, of all four colors. [Derived from ISO 12647-1]

NOTE 1 Sometimes known as the total dot area (TDA) or total area coverage (TAC).

NOTE 2 For most sets of color separation films the maximum of the tone value sum occurs at the position of the darkest achromatic tone of the image.

NOTE 3 The tone value sum may be determined from the color separation films or from the digital file.

Transmission densitometer   device which measures transmittance density. [ISO 12647-1]

Transmission density (or transmittance (optical) density)   logarithm to base 10 of the reciprocal of the transmittance factor. [ISO 5-2 and CIE Publication 17.4, 845-04-66]

Transmittance factor   ratio of the luminous flux transmitted through an aperture covered by a specimen to the luminous flux through the aperture without the specimen in place. [ISO 5-2]

EDITOR'S NOTE In obtaining the transmittance factor the value obtained will depend on the measurement geometry used, including the nature of the measurement aperture. Thus the measurement made may be the diffuse transmittance factor, the regular transmittance factor, or some combination of them both. However, it is common in densitometry and colorimetry to measure the diffuse transmittance factor, relative to the perfect transmitting diffuser as the reference.

Transparency (of an ink film)   the ability of an ink film to transmit and absorb light without scattering. [Derived from ISO 2846-1]

Transparency measurement values (of an ink film)   the reciprocal of the slope of the regression line between ink film thickness and color difference for overprints of chromatic inks over black. [Derived from ISO 2846-1]

Trapping   modification of boundaries of color areas to account for dimensional variations in the printing process by overprinting in selected colors at the boundaries between colors that might inadvertently be left uncolored due to normal variations of printing press registration. [ISO 15930]

NOTE This is alternatively referred to as chokes and spreads or grips and is not to be confused with the term "ink trapping."

Triplet comparison method   psychophysical method, defined in ISO 20462-2, which involves the simultaneous rank ordering of three test stimuli with respect to image quality or an attribute thereof, in accordance with a set of instructions given to the observer. [Derived from ISO 12231]

Tristimulus colorimeter   colorimeter which achieves the measurement values by the analog integration of the spectral product of object reflectance or transmittance factor, illuminant, and filters which are defined by the standard illuminant and standard observer functions. [Derived from ISO 12647-1]

Tristimulus value   amounts of the three reference color stimuli, in a given trichromatic system, required to match the color of the stimulus considered. [CIE Publication 17.4, 845-03-22]

Under color removal (UCR)   replacement of cyan, magenta, and yellow inks by black ink, in achromatic and near-achromatic colors only, such that the color is maintained.
NOTE This can be thought of as a special case of GCR.

Unicode   a character encoding standard which defines a unique number for every character in a large number of languages. The Unicode standard is maintained by the Unicode Technical Committee (http://www.unicode.org/standard/standard.html).

Variation tolerance   permissible difference between the OK print and that of a sample print taken at random from the production. [ISO 12647-1]

Veiling flare   relatively uniform but unwanted irradiation in the image plane of an optical system, caused by the scattering and reflection of a proportion of the radiation which enters the system through its normal entrance aperture. [ISO 12231]
NOTE 1 The veiling flare radiation may be from inside or outside the field of view of the system.
NOTE 2 Light leaks in an optical system housing can cause additional unwanted irradiation of the image plane. This irradiation may resemble veiling flare.

Veiling glare   light, reflected from an imaging medium, that has not been modulated by the means used to produce the image. [ISO 12231]
NOTE 1 Veiling glare lightens and reduces the contrast of the darker parts of an image.
NOTE 2 In CIE 122, the veiling glare of a CRT display is referred to as ambient flare.

Viewing flare   veiling glare that is observed in a viewing environment but not accounted for in radiometric measurements made using a prescribed measurement geometry. [ISO 12231]
NOTE The viewing flare is expressed as a percentage of the luminance of the adapted white.

White balance   adjustment of electronic still picture color channel gains or image processing so that radiation with relative spectral power distribution equal to that of the scene illumination source is rendered as a visual neutral. [ISO 12231]

Writer   application, system, or subsystem that generates a file based on predetermined criteria and prepares the file for output. [ISO 12639]

Part Two

Version 4

9

The Reasons for Changing to the v4 ICC Profile Format

The v2 ICC profile format specification was widely adopted by the color imaging community and proved very important in achieving and maintaining color fidelity of images across media, devices, and operating systems. This widespread use led to feedback from color management users and vendors that identified ways in which it could be further improved. That was the main driving force behind the v4 revision of the specification, which was first published in December 2001, and focused in particular on ways to improve interoperability.

Certain ambiguities in the previous versions of the specification occasionally permitted producers of profiles to misinterpret the reference color space and also the information they needed to provide in the profile. Thus profiles could be produced that were inconsistent with those produced by other vendors and when two such profiles were used together they could give rise to unexpected results. Furthermore, these ambiguities permitted ICC-compliant profiles to be produced that were interpreted slightly differently when used with different color management modules (CMMs). This meant that different CMMs could produce slightly different results to each other, even when using the same pair of profiles.

Although for many applications these problems were often small enough not to be an issue, there are other situations where high levels of consistency are particularly important. It was therefore necessary for the ICC to identify the major areas where ambiguities could permit poor interoperability and attempt to resolve those in the specification.

To understand the reasons for the main amendments to the specification it is helpful to put these in context. The changes are designed to ensure that profile builders understand the reference color space precisely, and exactly what is required of the profile. The changes also ensure that CMM producers are able to provide CMMs that ensure that any ICC-compliant profile is interpreted unambiguously by any ICC-compliant CMM, and that different CMMs processing the same pair of profiles to produce a color transformation provide a similar transformation. This improvement has largely been attained by removing ambiguities from the specification, rather than by imposing specific additional requirements on profile building or CMM developers – though there are some additional mandatory requirements.

Color Management: Understanding and Using ICC Profiles Edited by Phil Green
© 2010 John Wiley & Sons, Ltd

Thus this revision certainly does not mean that all profiles built for a specific device will be identical. There is still the need in many markets for profile building vendors to be able to differentiate their products and for users to select those products that best suit their needs. There is still no "one size fits all" in color reproduction and the ICC has not attempted to impose one. However, what it does mean is that when a user's preferred profiles are used, they should be produced in such a way that they are made to a common reference so that when combined with other profiles any results are predictable. This also means that when pairs of profiles are used, they should always produce the same result – regardless of which CMM is used. There is still a possibility that different CMMs could produce small differences due to differing interpolation procedures, but the more significant errors of interpretation have been removed.

Thus users will still need to select and build profiles that suit their reproduction needs – and ensure that they process the individual images to give their preferred reproduction within the context of those profiles. How this is done will be workflow dependent. The ICC is not proposing specific workflows and control procedures – that is the responsibility of the user and/ or specific industry standardization groups to recommend. However, within that context this version of the ICC specification provides users with the best tool for communicating the color rendering associated with devices to implement in their workflows.

Thus we can summarize the state of the art with this new specification as ensuring improved consistency when using ICC profiles. The system still retains the flexibility to let users produce profiles that best suit their requirements – they can choose when to trade off ease of use when building profiles against their individual needs. They can achieve this either by evaluating the various profile building software packages available and selecting the one that produces the best results for them, or by editing profiles to produce what they require. But because of the improved consistency, once a profile has been selected its performance in use should be highly predictable; and when pairs of profiles are used, they should always produce the same result, regardless of which CMM is used.

9.1 Summary of Changes

The changes made to the specification are summarized below. For details of the current specification, the full document is available at http://www.color.org/icc_specs2.html. Revisions agreed since the previous published version are listed at the same location.

9.2 Better PCS Definition

The job of the input profile is to define a transform from input device color values to the profile connection space (PCS). With v2 ICC profiles there were a number of different approaches taken to creating the perceptual rendering intent table for input profiles:

1. When creating scanner profiles, some profiling software simply adjusts the luminance range of color values from the scanned input medium. Since this is a scaling of luminance only, the range of color values presented by the input profile via the PCS to be the output profile differs substantially from one type of medium to another.
2. In other cases, profiling software maps image colors adjusted for a monitor directly into the PCS. Since the color gamut of a monitor is significantly different in shape from that of a

printer, an output profile that assumes colors have been mapped to a virtual print will clip many of the colors produced by this type of input profile.

3. Some digital camera profiling software attempts to estimate the colors in a scene. In some cases these colors are mapped to the PCS by a simple luminance scaling. As with approach 1, it is not possible to create an output profile that provides a good mapping for all scenes photographed because the range of colors presented to the PCS can vary significantly.

4. Other profiling software maps colors from input to an ideal reflection print as suggested in the specification. For v2 the ideal reflection print was poorly defined and so even in this case there is some variation in the mapping from one vendor to another.

This situation presents the output profile creator with a dilemma. It is possible to create an output profile that will provide an effective mapping for (say) an input profile for a transparency scan. This output profile will, however, produce a poor result when used with input profiles that perform different mappings to the PCS.

This problem is resolved for v4 profiles where a full definition of the perceptual PCS is provided along with the characteristics of the ideal reflection print used as its basis. The assumed level of illumination for viewing has also been specified For v4 the input profile must define a transform for the image from input color values to the ideal reflection print of the PCS. The output profile should provide a transform from the ideal reflection print to the output device. The v4 specification also indicates that the A2B table should provide as far as possible the reverse transform of the B2A table.

It should be noted that the change to the way in which the PCS in v4 is defined means that the v4 perceptual PCS and the colorimetric PCS are now different from one another.

9.3 Addition of the chromaticAdaptationTag

The PCS assumes that the ideal reflection print will be viewed in a standard D50 viewing environment. When measurements used to create the profile are made using a different illuminant, they must be adjusted using some form of chromatic adaptation. This is a common situation in the case where the input is monitor-like, where the measurement data is likely to be made relative to D65. In these situations a chromatic adaptation transform using a 3×3 matrix is usually performed to estimate equivalent colors under D50. Since this process involves estimating human perception of color (which is a complex process), there are a number of possible choices for this conversion, each of which produces a different result.

In some cases it is desirable to be able to recover the original measurement data, for example, in the case where a monitor-to-monitor color mapping is required. In order to be able to do this effectively, the chromatic adaptation transform used to map into D50 should also be used to map from D50.

In v2 profiles there is no way to determine which chromatic adaptation transform the profile creator used. In v4 information about the chromatic adaptation transform must be provided using the chromaticAdaptationTag, and the Bradford transform has been recommended as the default. When data is derived from, or intended for, viewing in illumination conditions other than those specified by ISO 3664 (i.e., D50), the transformation required for correction of the data must be specified.

This change is particularly important for color monitor profiles, which often do not assume a D50 chromatic adaptation state, but can have applications elsewhere (e.g., where prints or transparencies are expected to be viewed in non-standard conditions). An important consequence of this clarification is that v4 profiles for RGB displays and working spaces should only contain D50 tristimulus values in the mediaWhitePointTag indicating the transformation to the PCS white point.

9.4 Colorimetric Intents Are Required to Be Measurement Based

The definition of rendering intents has been made more precise to reduce ambiguities. In v2, profile builders were allowed to modify measurement data prior to building the relative colorimetric tables for a profile. This sometimes led to differences in the way in which colorimetric data could be interpreted when a colorimetric match is required.

The relative colorimetric rendering intent is now defined as measurement based. This requirement (together with the addition of the chromaticAdaptationTag and the improved media white definition) means that v4 ICC profiles can be used as the basis for "smart CMMs" where color conversions from the input to the output devices are calculated by the CMM at the time of output rather than at the time profiles are created. Since both input and output are known when the color transform is calculated, the result can be optimized.

9.5 Media White

The media white point specification has been improved. This ensures less ambiguity when calculating the absolute colorimetric rendering tables.

9.6 Unicode Support

There are a number of tags that hold human-readable descriptions. Version 4 introduces support for multi-byte fonts for these tags.

9.7 profileID

The addition of the profileID in v4 profiles assigns a more or less unique ID to each profile. This enables quick checking for identical profiles, and supports referencing profiles by their ID. (Version 2 profiles must be checked by comparing the entire file contents.)

9.8 Device N Color Support

The v2 specification allowed profiles with more than four channels; however, the colorant to be used is not defined for anything other than CMYK. This problem is solved for v4 profiles by the introduction of the colorantTableTag that defines the set of colorants by name and PCS color (i.e., their XYZ or $L^*a^*b^*$ coordinates).

9.9 Colorant Laydown Order

The v2 profile creators cannot indicate a difference in profiles made, for example, for print processes using the printing sequences CMYK and KCMY. This is a problem since the laydown order of inks changes the resulting color. This problem is solved for v4 profiles by the introduction of the colorantOrderTag that defines the laydown order of the inks.

9.10 Improved Color Processing Elements

Version 4 profiles allow the use of more capable look-up tables (LUTs) that provide applications developers with more flexibility, making it easier to define accurate color conversions. In addition, curves can be defined using parameters (parametric curves) rather than sample points, ensuring smoother color results. Another specification enables a simpler specification of one-dimensional LUTs for typical display devices.

These new LUT specifications overcome some issues of invertibility of the previous LUTs, as well as offering some other benefits of profile management by having a similar structure for all types of profiles.

9.11 Other Modifications

Clarifications have been introduced into the document covering such issues as the definition of the tags for three-component devices, the content and structure of monochrome profiles, the relationship between PCS XYZ and PCS $L^*a^*b^*$, and how to handle colors that can be represented in one and not the other.

Various new procedures have been specified to avoid confusion when using profiles, such as improved naming and dating procedures, and to permit profiles containing multiple rendering intents to be specified for input and display devices as they currently are for output profiles.

9.12 Approved Amendments

Since the v4 specification was first published, a number of amendments have been approved which further strengthen the interoperability and functionality of the profile format. These are described in more detail elsewhere in this book, and summarized in Chapter 5. The amendments are listed below with a brief outline of how they improve color management workflows.

9.12.1 Perceptual Intent Reference Medium Color Gamut

The PRMG amendment defines a virtual print medium with a large gamut that is recommended for use as the Perceptual Reference Medium with the v4 perceptual PCS. This enables source and destination profiles to connect with a known intermediate gamut; without such a common gamut the destination profile has to render from an unknown source gamut with serious consequences for interoperability.

9.12.2 Motion Picture Technology Tags

Tags indicating the use of input and output devices used in the motion picture industry extend the applicability of ICC profiles in this industry.

9.12.3 Floating Point Device Encoding Range

The Floating Point Device Encoding Range amendment introduces two important new features to the ICC architecture. First, it permits the use of floating point data, thus making it possible to extend the use of ICC profiles to applications where 8- and 16-bit integers have insufficient precision; and secondly it adds new flexible color processing elements in the BToDx and DToBx tags.

9.12.4 Profile Sequence Identifier Tag

By making it possible to identify the sequence of profiles used to generate an ICC DeviceLink profile, an image which has been prepared for one output-referred encoding can be converted to another output-referred encoding without ambiguity, for example, allowing the appropriate profile for the new encoding to be selected and embedded.

9.12.5 Colorimetric Intent Image State Tag

The CIIS tag extend the ICC architecture by making it possible for an input profile to be identified as representing scene-referred data, rather than the usual output-referred image state.

9.12.6 Deletion of media Black Point Tag

The mediaBlackPointTag was removed from the specification because there is no clear guidance on how it should be determined for given media and there is a lack of consistency in the way that vendors calculate and apply it. Where present in a profile, the mediaBlack-PointTag should now be considered a private tag.

9.12.7 Reasons to Adopt v4

Changes to the profile format introduced in v4 provide a number of advantages, the most significant of which follow from the removal of ambiguities from the specification and a more precise definition of the PCS. These lead to an improved predictability of performance of a profile in use which will lead to a reduction of major differences of interpretation. Therefore, when pairs of profiles are used, they should always produce the same result – regardless of which CMM is used.

Continued use of profiles in previous versions of the specification can cause color reproduction problems, including profile mismatches when a document is opened in an application set for v4. Workflows which treat v4 profiles as v2 may also lead to problems such as color shifts, when the v2 profile is a poor approximation of the v4 profile, and extra color conversions, when the v4 profile cannot be represented as v2 (e.g., sYCC).

Version 4 of the profile format has been adopted as ISO 15076. The ICC strongly recommends that vendors adopt this version of the profile specification. Further information, including implementation details for v4, are provided through the ICC web site http://www.color.org.

9.13 Mixing v2 and v4 Profiles

Many profiles in use today were constructed according to the v2 specification, and indeed there are still profile creation tools in use that generate v2 profiles. While the ICC recommends the use of the profile format defined in the v4 version of the specification, it also recognizes that the v2 format will remain in use in some workflows and thus the v2 specification will continue to be available through the ICC.

In moving to adoption of v4 it is not essential to discard all v2 profiles, since these profiles will, while retaining the ambiguities described above, interoperate with v4 profiles in a workflow.

Tests have shown that when using the media-relative colorimetric rendering intent, v4 source profiles combined with v2 output profiles give better consistency between different printers than a workflow where v2 source profiles are used in conjunction with v2 output profiles. If a v4 profile perceptual intent is used as the source and combined with v2 media-relative colorimetric for the output, with black point compensation turned on, this produces acceptable results since rendering is done on the source side as intended in a v4 workflow.

In v2 input profiles, it is usually only the media-relative colorimetric intent that is encoded in the profile, so when combined with a v2 output profile it is the perceptual rendering intent in the output profile that should be used.

For v2 workflows, tests indicate that a v2 source profile combined with a v2 output profile, with perceptual rendering on both source and output, produce the smallest color differences between source and output and hence may be an appropriate default when cross-printer consistency is not an issue.

It should be noted that when combining profiles, all current CMMs support both v2 and v4 profile formats, but not all applications support the use of different rendering intents for source and destination.

10

ICC Version 2 and Version 4 Display Profile Differences

10.1 Display White Point Adaptation

In Version 2 of the ICC specification, the assumed state of viewer adaptation to the display white point was not specified. Consequently, the chromatic adaptation applied to display white point tristimulus values to produce the PCS media white point values can range from no adaptation (the actual display white point values are encoded as the media white point) to full adaptation (D50 tristimulus values are encoded as the media white point). The result of this ambiguity is that different profiles for the same display can produce different results, depending on the degree of adaptation that was assumed by the profile maker. These differences can cascade through the rest of a color management workflow, as the appearance of images on a display is often used as a basis for color adjustments.

To resolve this ambiguity, Version 4 of the ICC specification requires that v4 display profiles assume the viewer is fully adapted to the display white point. This means that display tristimulus values must be chromatically adapted to the D50 PCS white point when creating the profile. However, the v4 specification also requires the chromatic adaptation matrix used to be included in the chromaticAdaptationTag if chromatic adaptation is needed (i.e., the display white point is not D50). This requirement makes it possible for CMMs to include the capability to undo the chromatic adaptation and obtain the actual display tristimulus values. Then, a capable CMM could reintroduce whatever degree of adaptation is desired.

Unfortunately, current CMMs do not offer a user-selectable degree of display chromatic adaptation. For most applications, this control is not necessary – fully adapted values produce the desired results. However, if some use case requires partial or no adaptation to the display white point, it may be necessary to use the appropriate v2 profile until such time as CMMs with chromatic adaptation control become available. This approach requires a high degree of knowledge and skill, and should only be employed by expert users. It is the belief of the ICC that the vast majority of user needs are met by assuming complete adaptation to the display white point, which is why this assumption was selected to remove the ambiguity.

Also, it appears that some color management users made use of profiles that assume no viewer adaptation to the display to modify the white point of the display without adjustment of the hardware. When using the relative colorimetric rendering intent, the display of the media white point of the source profile would be the white point set by the display hardware, but with the absolute colorimetric rendering intent the measured white point of an image seen on the display would be that of the media white point of the source profile – typically similar to D50. With v4 profiles there will be no such differences and so the only difference when the absolute or relative colorimetric rendering is used is that between the media white of the source profile and D50 itself.

Questions have been raised by some users as to how they can now obtain a white with the chromaticity of D50 on their display. If full adaptation is assumed to occur it should only be necessary to provide D50 on a display when direct comparisons are made to hard copy and full adaptation to the display is not possible. In such a situation users are recommended to follow the guidance of ISO 12646 and directly set the hardware to provide this chromaticity. Otherwise they are recommended to follow the guidance of ISO 3446 and set the hardware to provide the chromaticity of D65. However, if there are users who require D50 chromaticity, without resetting their hardware, their color management vendors should be encouraged to use the chromaticAdaptationTag to provide this functionality in the CMM.

10.2 Rendering Intents

ICC v2 display profiles typically contain only one rendering intent, and this rendering intent is typically a mixture of perceptual and colorimetric rendering. For example, most display profiles assume a display black point luminance of zero (no light whatsoever), and scale the measured display transfer function accordingly, but then otherwise encode display colorimetry in the profile. This approach results from two v2 characteristics: the perceptual intent black point is assumed to be scaled to zero; and there is no defined perceptual intent reference medium.

The problems with the above approach are as follows:

- Since real displays will not have a black point of zero, the display profile is not an accurate colorimetric profile. Furthermore, the scaling of the display black point will affect the encoded colorimetry of all the display colors except the display white point (even the display white point can be affected slightly by veiling glare).
- Since it is not possible to visualize and evaluate tone reproduction down to a luminance of zero, the ability to accurately view and control shadow detail is limited using v2 display profiles. Users can learn to compensate mentally for limited media in controlled situations, but this compensation is difficult to reliably communicate, or extend to arbitrary media.
- While in v4 there is a well-defined standard perceptual intent reference medium and associated gamut, there is no such medium defined for v2 and thus there is no way to color re-render the display colorimetry. The only opportunity for optimized color re-rendering is with proprietary situations where an output profile perceptual intent is tuned to receive the PCS colorimetry of a specific display profile.

The above issues result in limitations on the quality of images that can be produced using v2 display profiles. Historically, this has been less of an issue, because displays were less capable; users realized their limitations and performed final adjustments based on actual printed output. Also, scaling of media black point tristimulus values is a way to achieve reasonable first-order color re-rendering. However, as displays improve and display-based color encodings (such as sRGB) are widely used, it becomes important to know the true display colorimetry, to enable optimal quality color re-renderings to be produced.

The ICC v4 specification solves these problems, because it clarifies the inclusion of multiple rendering intents in display (and color space) profiles, and includes a well-defined perceptual intent reference medium with associated color gamut. Vendors of display profiling tools are encouraged to encode accurate display colorimetry in colorimetric intents, and to perform appropriate color re-rendering in perceptual intents where present. The art of color re-rendering is difficult to model mathematically to the extent required to create perceptual rendering intents automatically without user intervention. An example of a source-to-PRMG perceptual rendering can be found in the ICC sRGB v4 profile.

Many display profiling measurement devices do not record the ambient illumination which must be included to obtain accurate measurements of veiling glare. Accurate colorimetric intents are straightforward to construct using suitable measurement devices, and several manufacturers have shown profiles with high-quality (hand-tuned), display-to-print reference medium perceptual rendering intents.

11

Using the sRGB_v4_ICC_preference.icc Profile

11.1 Introduction

A major difference between v4 ICC profiles and earlier versions is the v4 Perceptual Reference Medium (PRM), which is a virtual large-gamut reflection print similar to prints obtained using high-quality photo printers and glossy paper. In the v4 specification, ICC introduced a perceptual PCS based on this reference medium. This is different from the colorimetric PCS which, like the v2 PCS, has a reference medium with a black point of zero and a white point corresponding to the perfect diffuse reflector under a D50 illuminant. While the colorimetric PCS is suited to reproduction goals where the source colorimetry is to be matched exactly to the destination colorimetry (within the limits of the destination color gamut), in most workflows the reproduction goal is to produce an optimal reproduction on the destination medium by mapping source white point to destination white point and applying any re-rendering needed to make full use of the output medium color gamut and compensate for differences in viewing conditions.

The intention when producing the v4 specification was to define a gamut for use with the PRM, so that when a perceptual conversion takes place both source and destination profiles use a common reference medium color gamut. This avoids the situation which occurred with the use of v2 profiles, where the output profile has to map from an unknown source color gamut and is unable to perform an optimal rendering for all possible source gamuts. In 2005 the ICC approved the Perceptual Reference Medium Gamut (PRMG), as defined in ISO 12640-3, as the recommended gamut for the perceptual PCS.

In implementing a v4 workflow, it is essential to have v4 profiles that do a good job of re-rendering between source encoding and the PRMG. Probably the most difficult task here is to generate a profile that renders between the sRGB encoding and the PRMG, since the gamuts are very different. To aid the process of v4 adoption, the ICC has provided a profile for this purpose.

Color Management: Understanding and Using ICC Profiles   Edited by Phil Green
© 2010 John Wiley & Sons, Ltd

The sRGB v4 ICC preference profile is a v4 replacement for commonly used sRGB v2 profiles, which gives better results in workflows that implement the ICC v4 specification.

The advantages of the new profile are:

1. More pleasing results for most images when combined with any correctly constructed v4 output profile using the perceptual rendering intent.
2. More consistently correct results among different CMMs using the ICC-absolute colorimetric rendering intent.
3. Higher color accuracy using the media-relative colorimetric intent.

Guidance on the use of this profile is given below.

11.2 Color Re-rendering and the v4 PRM

The v4 ICC profile perceptual intent transforms *color re-render* to and from the PRM in the ICC PCS, which is the color space where source and destination ICC profiles connect. Color re-rendering is the process where colors that are optimized for one medium are transformed to re-optimize them for a different medium. The transforms are designed considering the characteristics of the two different media, such as the dynamic ranges, color gamuts, viewing modes (e.g., monitor and print), and viewing environments. Color re-rendering is not needed when the characteristics of two media are similar, or when the goal is to produce an exact copy of the source medium on the destination medium (i.e., proofing). In the former case the perceptual and media-relative colorimetric transforms will be identical. In the latter case a colorimetric rendering intent should be used.

Good color re-rendering will maintain the artistic intent of the source image, and for the most part the appearance, although some colors will change as is necessary to deal with color gamut and viewing differences. Color re-rendering is to some extent image specific and a matter of personal preference, but it is possible to develop default color re-rendering transforms that produce results that are pleasing to most people when applied to most images. This was the design objective for the perceptual rendering intent of the sRGB v4 profile.

Color re-rendering transforms assume the source image is the intended reproduction for its medium, and will not attempt to correct or enhance a poor image, but will only re-optimize it for the destination medium. It is possible that in some cases a poor image will be slightly improved, but it is also possible that it will be made worse.

A side benefit of using color re-rendering transforms for cross-media conversions is that well-designed transforms like those in this profile tend to produce small errors when inverted. This means that they can be undone to enable re-purposing to different media with a small loss of color information (see roundtrip data below).

11.3 General Recommendations on Using
the sRGB_v4_ICC_preference.icc Profile

In workflows where only v4 ICC profiles are used:

- The ICC-absolute colorimetric rendering intent should be used when the goal is to maintain the colors of the original on the reproduction.

- The media-relative colorimetric intent should be used when the goal is to map the source medium white to the destination medium.
- The perceptual intent should be used when the goal is to re-optimize the source colors to produce a pleasing reproduction on the reproduction medium while essentially maintaining the "look" of the source image. The perceptual intent will not enhance or correct images.

CMMs may offer additional functions and rendering intents, such as:

- Black point compensation (BPC), where the source medium black point is mapped to the destination medium black point using CIE XYZ scaling.
- Partial or no chromatic adaptation instead of complete adaptation.

11.4 Differences between the sRGB_v4_ICC_preference Profile and v2 sRGB Profiles

The sRGB v4 profile is different from commonly used sRGB v2 ICC profiles in three fundamental ways:

1. The ICC-absolute and media-relative colorimetric rendering intent transforms are not black point scaled.
2. The ICC-absolute colorimetric transforms are correct implementations of the ICC v4 specification, which has been defined in a narrower way than the ICC v2 specification and assumes the viewer is fully adapted to the display white point.
3. The perceptual rendering intent transforms use the v4 PRM assuming the PRMG.

These differences will produce different results when the v4 sRGB profile is used, as compared to commonly used v2 sRGB profiles.

It should be noted here that the PRM is defined in all ICC profile format specifications starting with ICC.1:2001–12 and in ISO 15076-1. The PRMG is defined in the approved amendment to the ICC v4 specification, and is incorporated in ISO 15076-1:2010.

11.5 ICC-Absolute Colorimetric Rendering Intent

ICC v2 sRGB profiles can be grouped into five different types depending on the nature of the ICC-absolute colorimetric rendering intent:

1. Those that assume full adaptation to the display white point and do not include BPC.
2. Those that assume full adaptation to the display white point and include BPC.
3. Those that assume partial or no adaptation to the display white point and do not include BPC.
4. Those that assume partial or no adaptation to the display white point and include BPC.
5. Profiles where the RGB to XYZ matrix assumes full adaptation, but the media white point values are unadapted. Such profiles typically include BPC (but do not have to).

ICC v2 sRGB profiles of types 1 and 2 can be downloaded from http://www.color.org/srgbprofiles.html.

ICC v2 profiles of types 3 and 4 are uncommon, but can be constructed for special purposes.

Unfortunately, the widely used *sRGB Color Space Profile.icm* (sRGB IEC 61966-2.1) v2 profile is type 5.

The ICC-absolute colorimetric intent of the new ICC v4 profile is comparable to a type 1 sRGB v2 profile. Due to clarifications of the ICC specification, type 2, 4, and 5 profiles are not valid as v4 profiles. The additional restrictions are intended to improve accuracy and reduce the variations of results achieved from using the ICC-absolute colorimetric rendering intent with different CMMs. Details are provided below.

11.6 Results Using the ICC-Absolute Colorimetric Rendering Intent of the v4 sRGB Profile Versus Using a v2 sRGB Profile

When using the ICC-absolute colorimetric intent of the v4 sRGB profile, the following results should be obtained when compared to using a v2 sRGB profile:

- The results using type 1 v2 sRGB profiles should be the same as when using the v4 sRGB profile. In this case, the only differences will be due to the precision of the profiles. The v4 sRGB profile uses a mathematical function instead of a LUT.
- The results using type 2 v2 sRGB profiles should be the same as when using the v4 sRGB profile with BPC (but many CMMs do not enable BPC in combination with the ICC-absolute colorimetric rendering intent).
- The results using type 3 v2 sRGB profiles can be achieved using the v4 sRGB profile in combination with a CMM that supports partial or no adaptation (but these are rare).
- The results using type 4 v2 sRGB profiles can be achieved using the v4 sRGB profile in combination with a CMM that supports partial or no adaptation and BPC.
- Type 5 v2 sRGB profiles are internally inconsistent, and will not produce correct results without a case-specific correction applied by the CMM. Some CMMs will fix the profile while others will not. Thus, the absolute colorimetric rendering intent of a type 5 v2 sRGB profile will produce different results depending on the CMM used. CMMs that fix type 5 v2 profiles then often produce incorrect results with type 3 and type 4 v2 profiles.

11.7 Media-Relative Colorimetric Rendering Intent

Using this rendering intent:

1. The v4 sRGB profile will produce approximately the same results as a type 1 v2 sRGB profile.
2. The v4 sRGB profile should produce the same results as a type 2 or type 5 v2 sRGB profile if the CMM uses the relative colorimetric rendering intent and BPC:

(a) It is not possible to obtain results equivalent to those obtained with a v4 profile and BPC off using a type 2, 4, or 5 profile. If BPC is undesired, either a v4 sRGB profile or a v2 profile without BPC should be used.

(b) The only way to produce the same results using the v4 profile as obtained using a v2 profile with BPC included in the profile but with BPC off in the CMM is if the CMM enables BPC to be on for the v4 sRGB profile but off for the other profile. Typically this option is not available. Thus, if this result is desired it may be necessary to use a v2 sRGB profile with BPC included.

(c) The only way to produce the same results using the v4 profile as obtained using a v2 profile with partial or no adaptation to the display white is to use a CMM that supports partial or no adaptation.

11.8 Perceptual Rendering Intent

The perceptual transforms in the v4 sRGB ICC preference profile are bidirectional, providing color re-rendering from sRGB to the PRM when the profile is used as a source profile, and providing color re-rendering from the PRM to sRGB when the profile is used as a destination profile. The roundtrip errors are larger than for the colorimetric intents, but are still small. For the sRGB $\rightarrow$ LAB $\rightarrow$ sRGB roundtrip, for all 8-bit RGB code values:

$$\text{Perceptual mean 8-bit RGB code value error, mean } \Delta\text{RGB} = 0.225$$

$$\text{Perceptual maximum 8-bit RGB code value error, max } \Delta\text{RGB} = 3.28.$$

For the LAB $\rightarrow$ sRGB $\rightarrow$ LAB roundtrip, for 1168 color patches that are on a $19 \times 19 \times 19$ uniform grid and inside the AtoB0 gamut, the results are:

$$\text{Perceptual mean } \Delta E = 0.27$$

$$\text{Perceptual maximum } \Delta E = 4.20.$$

Using the perceptual rendering intent, all the primaries and secondaries (red, green, blue, cyan, magenta, yellow, white, black) invert perfectly using 8-bit encoding:

$$\text{RGB} = \text{RGB} \rightarrow \text{PCS} \rightarrow \text{RGB}$$

Generally, the perceptual intent of the v4 sRGB ICC preference profile should only be used with the perceptual intent transforms of other v4 profiles, as such transforms are required to also color re-render to and from the PRM. It is best to use v4 profiles that indicate the use of the PRMG through the perceptualRenderingIntentGamutTag (the rig0 tag is set to PRMG), for maximum interoperability.

If the v4 sRGB ICC preference profile is embedded as the source profile, and it is necessary to use it with a v2 destination profile, and the intention is to use a perceptual rendering intent, then there are several options:

1. If the objective is to produce a large-gamut photo print, convert using the perceptual rendering intent to a color encoding such as ROMM RGB that uses the ICC PRM as its reference medium, and for which a v4 profile is available. Then convert from ROMM RGB to the destination device encoding using media-relative colorimetric with BPC on.
2. Use the media-relative colorimetric rendering intent with BPC as a baseline perceptual color re-rendering to convert directly from sRGB to the destination device encoding.
3. Temporarily replace the v4 sRGB profile with a type 2 v2 sRGB profile (full adaptation to display white point and BPC included) and use the v2 perceptual intent.

In most cases these options will produce acceptable results, but they may be different from the results that would be obtained if a v4 destination profile were used.

Likewise, the v4 sRGB ICC preference profile should generally not be used as the destination profile with v2 source profiles. An exception to this is when the source image colorimetry is for a medium similar to the PRM, such as a large-gamut photo print. In this case the v2 source profile can be used to convert to a color encoding that uses the PRM (such as ROMM RGB) using the media-relative colorimetric rendering intent with BPC on. Then, the v4 perceptual rendering intent is used to color re-render from the encoded print colorimetry to sRGB.

The v4 sRGB ICC preference profile was developed and tested by ICC members. The description tag of this profile currently contains the content "sRGB v4 ICC preference perceptual intent beta" to indicate that the perceptual intent contains a preference re-rendering from sRGB to the PRMG and vice versa, and to further indicate that it is currently in a beta state and that users are encouraged to provide feedback. If no significant complaints are reported the extension "beta" will be removed from the description tag.

The v4 sRGB ICC preference profile can be downloaded from http://www.color.org/srgbprofiles.html.

11.9 Notes on Workflow

This section contains detailed descriptions of four different example workflows.

11.9.1 Re-rendering from sRGB to Output Medium Gamut

When the v4 sRGB profile is assigned to sRGB images and used as the source profile, the perceptual rendering intent is designed to transform image colors optimized for sRGB displays into image colors optimized for the PRM. A v4 destination printer profile is intended to be used in combination with the v4 sRGB profile to produce colors on actual print reproductions (see Figure 11.1).

11.9.2 Re-rendering from sRGB to PRMG with Subsequent Colorimetric Proof

In the case of a large-gamut printer, the v4 sRGB profile perceptual intent can also be combined with a colorimetric rendering intent in the destination printer profile (see Figure 11.2). This will result in a "proof" of the perceptual intent reference medium colorimetry.

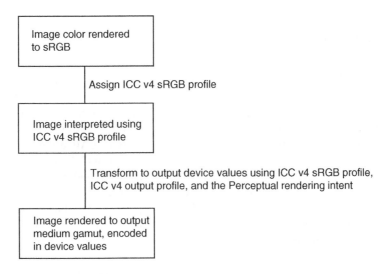

Figure 11.1   Using the v4 sRGB profile with a v4 output profile and the perceptual rendering intent

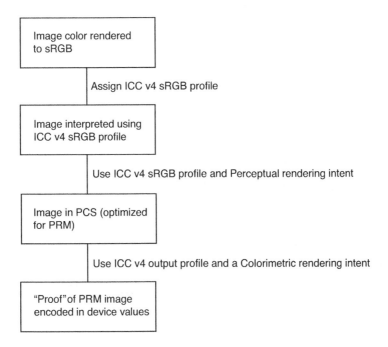

Figure 11.2   Using the v4 sRGB profile perceptual rendering intent, a v4 output profile, and the colorimetric rendering intent

NOTE: BPC may also be used with the relative colorimetric rendering intent to perform a simple color re-rendering from the PRM to the output device.

11.9.3 Re-rendering from sRGB to Intermediate Color Space with Subsequent Colorimetric Proof

With Adobe Photoshop, a two-step process must be employed to use different rendering intents for source and destination. First the sRGB image is transformed using the perceptual intent to an intermediate color space that is appropriate for the PRM colorimetry, and then the result is transformed using a colorimetric intent to the destination color space (see Figure 11.3).

It should be noted that if the intermediate color space used is the Photoshop LAB color space, BPC will be applied automatically when going from sRGB to LAB using the perceptual rendering intent. Consequently, it must also be used when going from LAB to the device values in order to map the sRGB black point, as re-rendered to the PRM and then scaled to zero in LAB, to the black point of the device. This is necessary because v2 profile perceptual intents typically include black point scaling to zero, and Photoshop uses a v2 profile for the LAB color space. However, if an intermediate color space based on a v4 profile is used (such as ROMM RGB) the PRM black point is left unchanged, and when going to the actual output medium BPC can be either on or off, as desired.

A color space is suitable for use as an intermediate color space in this process if the color space profile absolute colorimetric intent does not perform any color rendering or re-rendering, and does not clip or otherwise alter any colors within the PRMG.

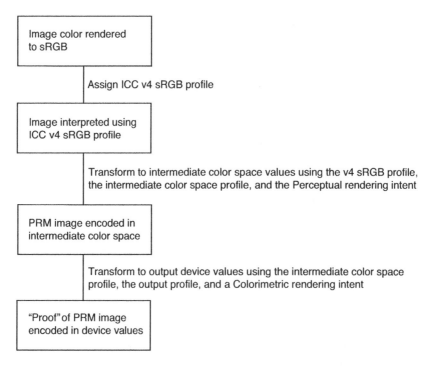

Figure 11.3 Using the v4 sRGB profile perceptual rendering intent, a v4 output profile, and a colorimetric rendering intent in a two-step process

11.9.4 Re-rendering Print-Referred Data to sRGB

The ICC v4 sRGB profile can also be used as a destination profile to produce sRGB images from images containing reflection print colorimetry encoded as ROMM RGB, or LAB (such as the ISO 12640-3 SCID images), or printer device values with an appropriate ICC v4 profile (see Figure 11.4).

Since the ICC v4 sRGB profile performs a color re-rendering in both directions (with no clipping inside the PRMG), there is minimal loss when converting from the PRM to sRGB and back again. This makes it possible to communicate large-gamut print colorimetry encoded as sRGB if the v4 sRGB profile is embedded and the perceptual rendering intent is indicated. The intended colorimetry for reproductions on specific media can also be communicated by embedding both the v4 sRGB profile and the output intent profile, such as by using PDF/X-4.

Because of interpolation issues, multiple roundtrips from sRGB to the PRM and back can result in errors accumulating to the point of significance, so the v4 sRGB profile perceptual intent should not be used as a print-referred working space if multiple roundtrips are anticipated. In this case ROMM RGB is a better choice, with the v4 sRGB profile used initially to convert to the working space, and finally to convert from the working space to sRGB after all edits and adjustments are completed.

Inferior results may be obtained if Adobe RGB images are converted colorimetrically to sRGB and then the ICC v4 sRGB profile is applied, because the colorimetric conversion from Adobe RGB to sRGB may not produce optimal sRGB colorimetry. As a general rule, if Adobe RGB images are optimized based on previewing them on an sRGB display, it will likely be possible to convert them colorimetrically (with clipping) to sRGB and assign the v4 sRGB profile. For Adobe RGB images optimized by viewing prints produced using large-gamut photo printers and a colorimetric rendering intent, acceptable results may be

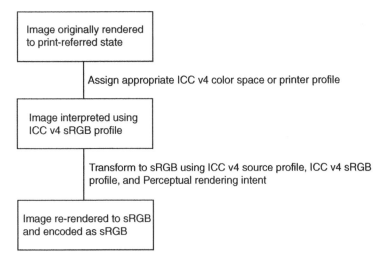

Figure 11.4   Using a v4 source profile, the v4 sRGB profile, and the perceptual rendering intent to re-render print-referred data to sRGB

obtained by converting from Adobe RGB to sRGB using the Adobe RGB profile as the source profile and choosing the colorimetric rendering intent used to make the prints, and the v4 sRGB profile as the destination profile and choosing the perceptual rendering intent to sRGB, as illustrated in Figure 11.4. This may require a two-step process if the software used does not support the selection of different rendering intents for source and destination.

12

Fundamentals of the Version 4 Perceptual Rendering Intent

ICC Version 4 differentiates clearly between perceptual rendering and colorimetric rendering so that the applications appropriate for each of these rendering intents are clarified. Improved workflows can be achieved by exploiting these definitions of clarified rendering intent.

An understanding of image state concepts will assist in understanding and applying the ICC perceptual rendering intent. (A definition of image state can be found in ISO 22028-1 [1].) Essentially, the image state conveys information content potential pertaining to encoded color information. As color scientists we know that scenes in general have certain extents of color and tone information, scanned hard copy originals in general have certain different extents of color and tone information, and so on. From this general understanding, the image state semantic allows us to categorize encoded color information – based on real-world algorithm and encoding capabilities and constraints. A color object encoded in a particular image state is appropriate for the uses and output modes associated with that image state. Furthermore, the concept of image state allows us to clarify our understanding of the image processing relationships between different color information content potentials – that is, between different image states, for example, the fundamental processing required when transforming a scene to an image suitable for reflection print output.

In general, recently developed color image encodings are each identified with a particular image state, with an associated color space white point, and viewing environment. A color gamut, with a particular volume and luminance range, can be a part of a particular image state condition. Note, however, that while, in a sense, image state is an attribute of a color image encoding, an image state is in fact a representation of what can be done with any color object encoded for that image state. Several image encodings are valid for use with each of the standardized image states: scene referred, original referred, reference output referred, and actual output referred.

With these image state concepts in mind, the ICC perceptual rendering intent can be defined. This perceptual rendering intent is provided to accomplish a *preferential* adjustment in concert with an image state–image processing transition.

Color Management: Understanding and Using ICC Profiles   Edited by Phil Green
© 2010 John Wiley & Sons, Ltd

A comparative look at the colorimetric rendering intents can help to further position the perceptual rendering intent. The media-relative and absolute colorimetric rendering intents provide a means to transition from one color space encoding to another, adapting for color space white point differences while maintaining colorimetric measurement accuracy for in-gamut colors. Image data is re-encoded, via any of the colorimetric renderings, but is not adjusted preferentially for image state differences. The only image state constraints that are incorporated via colorimetric renderings are gamut volume (when a particular gamut volume is associated with the target image state condition) and color space white point. Essentially, either of the colorimetric intents can be used to re-encode image data, *while maintaining a current image state*, for example, capture referred, output referred. In addition, either of the colorimetric intents may be appropriate for transitioning between two closely related image states, such as reference output referred (e.g., ICC PCS reference medium) and actual output referred, for example, when the actual output condition is similar to that of the reference output condition.

The distinction in the perceptual rendering intent is now explained: it provides a means to transition from one image state to another image state, preferentially adjusting color appearance for differences in any or all image state characteristics. In transition, colors are adapted to achieve a preferred color appearance within reference or device constraints, and out-of-gamut colors that cannot be represented in the destination image state are adjusted using one of many gamut mapping strategies. Note that if a reference output-referred and an actual output-referred image state are essentially identical, then a perceptual rendering intent transforming between those states can be thought of as performing a NULL image state transition. In this case the perceptual intent can be identical or similar to a media-relative colorimetric intent.

Given this background, one understands that the preferential nature of any *particular* perceptual rendering intent is image state transition dependent. For example, the preferential nature of a perceptual rendering intent used to transition from a raw digital camera RGB to ICC PCS should be different from the preferential nature of a perceptual rendering intent used to transition from ICC PCS to a printer CMYK. The image state transition from raw digital camera RGB to ICC PCS reference medium is scene referred to output referred (reference). (*Note that this initial image processing from scene referred to output referred occurs inside almost all digital cameras – the image written from the camera is output referred.*) The image state transition from ICC PCS reference medium to a printer CMYK is output referred (reference) to output referred (actual device constrained). One part of the difference between a "scene-referred to output-referred transition" and an "output^B-referred to output^A-referred transition" is that color rendering from a natural scene to an image requires specific preferential handling, adapting the color information from the three-dimensional world to the two-dimensional imaging environment.

Given that a perceptual rendering intent transform applies a preference adjustment, a perceptual rendering can be understood to target a particular image state color appearance, that is, "color aim." A color aim is the color appearance goal of a preference adjustment or adaptation. A color appearance "color aim," dependent on source and destination image states, is inherent in all ICC perceptual rendering intent transforms. However, due to the nature of ICC profiles, the inherent color aim in perceptual rendering intent transforms is not visible to or tunable by the users of ICC profiles.

Color rendering of scenes (scene-referred image state) to create reproductions (output-referred image state) typically includes a chroma and contrast boost. This is an example of an image state appearance preference adjustment. This boost must be done only in the device-to-PCS perceptual transform of an input (scene-referred to output-referred) ICC profile. This boost is by nature a non-convergent operation; that is, if it is applied repeatedly it produces unacceptable results. The output-referred image state of the ICC PCS perceptual intent reference medium serves as a target for this scene-referred to output-referred perceptual color rendering. Output^B-referred to output^A-referred ICC PCS-to-device perceptual transforms (e.g., perceptual rendering intent transforms in printer profiles) should not implement this particular chroma and contrast boost.

For general purpose pictorial reproduction, perceptual rendering intent transforms are applied in both the input to ICC PCS (scene-referred to output-referred) and ICC PCS to printer (output^B-referred to output^A-referred) image state transitions. When a perceptual rendering intent transform has been used to color-render into ICC PCS, the intermediate ICC PCS "image" is the media-relative colorimetric (reference medium output-referred) re-presentation of an idealized color appearance visualization appropriate to the constraints of the reference medium. In ISO 22028-1 terms, ICC PCS is a color space encoding and the perceptual rendering intent result in ICC PCS is a color image encoding. The general purpose pictorial reproduction is completed when the ICC PCS color image encoding is perceptually color re-rendered to an actual visualization (actual output referred).

Alternatively, in cases when the digitization (capture) goal is to accurately retain the image state of a limited gamut source image (e.g., is the source image gamut $\sim$288:1 linear dynamic range from a reflection print scan?), media-relative colorimetric rendering from capture to ICC PCS can be followed by perceptual (capture-referred to output-referred image state transition) or media-relative colorimetric (capture image state is essentially preserved) rendering to visualization. In this case ICC PCS holds capture-referred, media-relative colorimetric values. Preferential image state transition-dependent adjustments to output conditions (capture referred to output referred) are handled through the output profile. Note that this places a particular constraint on the "color aim" to be achieved in the output profile ICC PCS-to-device perceptual rendering intent transform. Media-relative colorimetric intents may be appropriate for each of the encoding transitions from original reflection print digitization to reproduction printing, given that the information is consistently related to reflection print color capability.

In any ICC PCS-to-device transition, resulting in an actual output-referred image state, the selection of perceptual rendering intent versus one of the colorimetric rendering intents must take into account the image state of the image in ICC PCS (e.g., how was the image "encoded" into ICC PCS?) and the similarities and differences between that ICC PCS image state and the targeted actual output-referred image state. The differences and similarities are judged in terms of the image state attributes: color space encoding, color space white point, viewing environment, appearance aim relative to a reference medium, and color space gamut – having a particular volume shape and luminance range.

The v4 ICC PCS defines the dynamic range of the perceptual intent reference medium, and also suggests that the reference color gamut defined in Annex B of ISO 12640-3 [2] is used to define the Perceptual Reference Medium Gamut. The PRMG approximates the maximum gamut of real surface colors, and using it as the rendering target of the perceptual intent assures that colors that have been rendered to the PCS are consistently defined. This eliminates the need for re-rendering by the output profile perceptual rendering intent.

References

[1] ISO (2004) 22028-1:2004. *Photography and graphic technology – Extended colour encodings for digital image storage, manipulation and interchange – Part 1: Architecture and requirements.* International Organization for Standardization, Geneva.
[2] ISO (2007) 12640-3:2007. *Graphic technology – Prepress digital data exchange – Part 3: CIELAB standard colour image data (CIELAB/SCID).* International Organization for Standardization, Geneva.

13

Perceptual Rendering Intent Use Case Issues

The perceptual rendering intent is used when a pleasing pictorial color output is desired. This differentiates it from a colorimetric rendering intent, which is used when an output is to be color matched to its source image. The perceptual rendering intent is most often used to render photographs of scenes (i.e., views of the three-dimensional world), and when the objective for a reproduction is to obtain the most attractive result on some medium that is different from the original (i.e., re-purposing), rather than to represent the original on the new medium (i.e., as in proofing or re-targeting). Some level of color consistency is usually required – for example, colors should not change hue names. However, with perceptual rendering, if the reproduction medium, for example, allows for greater chroma than the original medium, then chroma may be increased to produce a more pleasing result. Likewise, if the reproduction medium has a smaller color gamut than the original medium, perceptual rendering may alter in-gamut colors to allow for graceful accommodation of the original color gamut through gamut compression. In comparison, colorimetric rendering maintains in-gamut colors across media at the expense of suboptimal colorfulness on larger gamut reproduction media and clipping artifacts on smaller gamut reproduction media.

Keep in mind that the perceptual rendering intents in ICC profiles provide one approach to perceptual color rendering or re-rendering. There are other ways. Devices such as digital cameras and printers perform embedded (typically proprietary) perceptual renderings to and from standard color encodings like sRGB. In certain workflows, abstract ICC profiles can be used in combination with a colorimetric rendering path through source and destination ICC profiles to perform color re-rendering from source image colorimetry to destination image colorimetry directly in the PCS, before transforming to the destination encoding. Alternatively, a user may apply manual image editing techniques to optimize an image for a particular output condition. Finally, a color management system (CMS) may offer color rendering or re-rendering capabilities beyond that built into any source and destination profiles.

"Media-relative colorimetric plus black point compensation" is a simple and widely used perceptual rendering that uses the media-relative colorimetric rendering intent in the source and destination ICC profiles, combined with black point scaling performed by the CMS. Simple media white and black scaling can accommodate differences in dynamic range between an original and a reproduction and (to some extent) differences in color gamut size. In cases where color gamut shapes are roughly similar, and gamut size differences correlate with white and black point differences, media-relative colorimetric plus black point compensation may produce excellent perceptual rendering. However, this approach is not universally available because some CMSs do not support black point compensation. In other cases, more elaborate perceptual transforms are required to produce optimal results, especially when the source and destination media are quite different. The inclusion of an explicit perceptual rendering intent in ICC profiles enables well-defined, repeatable, and high-quality perceptual rendering across all ICC-based CMSs.

13.1 Scene to Reproduction

Scene-to-reproduction perceptual rendering is discussed first because such color rendering must happen in the capture of natural scenes, and understanding this transformation is helpful in understanding subsequent transformation requirements. However, users should be aware that in typical digital camera workflows, scene-to-reproduction perceptual rendering is not accessible to user control. Virtually all digital cameras perform scene-to-reproduction color rendering in the camera. The image file output by the camera does not represent the scene, but rather represents what the camera manufacturer feels will likely be a pleasing reproduction of the scene. This reproduction typically includes alterations of the scene colorimetry, including highlight compression, and mid-tone contrast and colorfulness enhancements as discussed below.

Likewise, camera raw processing applications typically embed scene-to-reproduction color rendering. While it is possible to create true scene-referred images from camera raw image data, most camera raw processing applications do not support this. Camera profiling applications include scene-to-PCS color rendering but may not offer user controls (note that with some camera profiling applications the accuracy of the scene color analysis is limited more by the accuracy of the target-based characterization method than by intentional preferential alterations).

In the future, it is expected that users will have more access to scene-referred image data, thereby gaining more explicit control over scene-to-reproduction color rendering. At present, these paragraphs are included primarily as background, and for an understanding of custom workflows where special camera modes or processing applications are used to enable true scene-referred image creation, followed by scene-to-reproduction color rendering.

At this point, the reader who is not familiar with image state concepts may wish to refer to the definitions and discussion of image state in ISO 22028-1 [1]. The ICC perceptual rendering intent operates intrinsically as an image state transition mechanism and the discussion that follows uses that terminology. The image state indicates how the encoded color information is to be interpreted. Scenes in general have different extents of color and tone information than scanned hard copy. From this general understanding, the image state semantic allows us to categorize encoded color information – based on real-world algorithm and encoding capabilities and constraints. A color object encoded in a particular image state is appropriate for the uses and output modes associated with that image state. Furthermore, the concept of image state allows us to clarify our understanding of the image processing relationships between different color information content potentials – that is, between different image states, for example, the fundamental processing required when transforming a scene to an image suitable for reflection print output.

An ICC profile is typically understood as associated with a device condition or a workspace color encoding. In fact, the perceptual rendering intent transform within an ICC profile is also tuned to accomplish a particular image state transition. With this in mind, we understand that ICC profiles are device condition – and image state condition – specific.

The essential process in any scene-to-reproduction (scene-referred to reference output-referred transition) perceptual transformation is a coordinated combination of color appearance adaptation, preference adjustments, and gamut mapping. This perceptual rendering intent *color rendering* transformation is used to map scenes to the fixed range of a reproduction in a pleasing way (where the term "color rendering" explicitly connotes that an image state transition is included in the color processing transformation). When a source image is scene referred, the device-to-PCS perceptual transform performs a perceptual rendering from the scene to the perceptual intent reference medium. Note that in an ICC v4-compliant (scene-referred) input profile (e.g., a digital camera input profile), the reference output-referred to scene-referred PCS-to-device perceptual rendering intent transform should invert (i.e., undo) that profile's own device-to-PCS perceptual rendering intent transform.

Commonly, the color appearance adaptation portion of a perceptual color rendering transformation includes adaptation from the scene adopted white (both the chromaticity and luminance) to the adopted white of the reproduction. Reproduction constraints and color appearance preferences determine the mapping of the adopted white, adapted scene colorimetry to produce a pleasing reproduction. For example, if the scene luminances are much higher than those of the reproduction in the anticipated viewing conditions, a chroma boost may be necessary to maintain the appropriate colorfulness. The anticipated surround of the reproduction can affect the desired contrast, with darker surrounds requiring higher contrast. Preferences play a significant role in determining this mapping, as viewers tend to prefer increased colorfulness and contrast in reproductions, to the extent that the increases do not look unnatural. Ideally, mappings are determined on a scene- and output medium-specific basis, implying image-specific perceptual intents. In production workflows fixed mappings that work reasonably well for most scenes are often used. These mappings typically boost the scene gamma and mid-tone contrast. For example, film reproduction systems have a mid-tone gamma greater than unity ($\sim$1.2–1.6, depending on the anticipated output medium) combined with highlight and shadow roll-offs. This s-shaped mapping allows film systems to accept both low and high dynamic range scenes, while maintaining preferred mid-tone contrast and colorfulness. Likewise, video systems have a system gamma of $\sim$1.2–1.4 and some highlight compression (at least in high-end systems).

The preference adjustment portion of a perceptual color rendering transformation often includes preferential expansion or compression of the source gamut and dynamic range to match that of a particular output (visualization) medium. Source scene gamut expansion and compression may be determined based on the *potential* scene extent from a particular digitization source device. Alternatively, in scene-specific color rendering cases, the extent of each specific source scene gamut may be evaluated and preferentially expanded or compressed to match the output medium. In some cases, preferential mappings also explicitly consider the reproduction of memory colors. Following such appearance–preference mapping, it may be necessary to apply gamut mapping to bring the remapped colors to within the actual gamut of the destination medium. Ideally the appearance–preference mapping would accomplish this, but practically, a following gamut mapping operation may be required. Note that the perceptual rendering intent color rendering provided in v4 input profiles targets the ICC perceptual intent reference medium.

Optimal preference mappings differ for scenes of low, medium, and high dynamic range, key, and gamut extent. Some scenes have colors out to the spectral locus (and beyond, after chromatic adaptation) and have very high luminance (dynamic) ranges; however, many scenes do not. In fact, most scenes have dynamic ranges (and gamuts) smaller than the 288:1 of the ICC perceptual intent reference medium. ICC profiles are typically used in capture condition or visualization condition (i.e., image state) specific – rather than image-specific – workflows. With these workflows, customizing the choice of rendering intent is one way to adapt the use of an ICC profile to a particular scene or color object.

It should be noted that the capture digitization of an original (two-dimensional) artwork or photograph (original-referred image state) is different from the capture of a scene, which is a view of the natural (three-dimensional) world. The discussion above relates to the capture of scenes. The capture of originals, even using a digital camera, falls under re-targeting or re-purposing as discussed below. Perceptual rendering intents for scene capture will generally not be appropriate for the capture of two-dimensional originals.

13.2 Re-targeting and Re-purposing

After data is color rendered to a particular reference output-referred or actual output-referred *first visualization* condition, that is, output-referred image state, it may be necessary to transform the data for a *second visualization*. For example, in a typical digital camera workflow, the "pleasing reproduction of the scene" produced by the camera is targeted for viewing on a soft copy display. That display-referred data may be color re-rendered when a print output is desired. Two scenarios are defined regarding such color transformations. When the second visualization is intended to represent or match the original first visualization, this is called *re-targeting*. Re-targeting is typical for "proofing." When the second visualization is independent of (i.e., not constrained by) the first visualization and can be optimized for the second visualization condition, this is called *re-purposing*. Keep in mind that both re-targeting and re-purposing are intended to operate on source images that are already in a picture-referred image state (either original or output referred, but not scene referred).

In re-targeting, the device-to-PCS media-relative colorimetric transform of the first visualization output or display profile is sequenced with the PCS-to-device media-relative colorimetric transform of a second visualization output or display profile. (Absolute colorimetric intents can be used when the color of the target substrate from the first visualization is to be carried through to the second visualization.) No new or revised image state preferential rendering is called for in re-targeting. The accuracy of the representation through the second visualization condition will be proportional to the capability of the second visualization condition to match the first visualization condition (e.g., gamut volume shape, luminance range, and color differentiation).

In re-purposing, the first concern is to remove the constraints in the color data that were induced by the prior perceptual rendering for a particular visualization condition (constraints preferentially based on a color aim determined as a function of prior source and destination image states). It is problematic that the constraints induced by a first preferential color rendering cannot be determined by examining color data after it has been so rendered. Color aim preferential rendering behavior is also not easily determined by examining the perceptual rendering intent transform of an output profile. Further, preferential capabilities in a CMS may have contributed to the first visualization, and can be difficult to extract in preparation for a later visualization.

In support of re-purposing, the ICC v4 specification places a new emphasis on perceptual rendering intent transformations:

- In ICC v4-compliant (actual output-referred) output profiles, the actual output-referred to reference output-referred device-to-PCS perceptual rendering intent transform should invert (i.e., undo) that profile's own PCS-to-device perceptual rendering intent transform, to allow for re-purposing from the ICC perceptual intent reference medium.
- In ICC v4-compliant (original-referred) color space encoding profiles and scanner profiles (e.g., an sRGB profile, document scanner input profiles), the device-to-PCS perceptual rendering intent transform should color re-render the original to an appropriate ICC perceptual intent reference medium representation (i.e., transform from the device, or encoding, medium image state to the ICC perceptual intent reference medium image state).
- In ICC v4-compliant (original-referred) color space encoding profiles and scanner profiles (e.g., an sRGB profile, document scanner input profiles), the PCS-to-device perceptual rendering intent transform should color re-render back to the original (i.e., transform from the ICC perceptual intent reference medium image state to the device, or encoding, medium image state) to allow for a new re-purposing directly from the original-referred image state. Note that in order to provide for a lossless roundtrip, this PCS-to-device perceptual rendering intent transform should be an inverse of the device-to-PCS perceptual rendering intent transform.

When transforming to the ICC perceptual intent reference medium image state, a reference color gamut should form part of the rendering target, as well as the fixed perceptual intent PCS dynamic range defined in v4 of the ICC specification. The ICC recommends that the color gamut defined in Annex B of ISO 12640-3 is used as the PRMG. Media-relative CIELAB L^*, C^*, and h_{ab} values for the boundary of this gamut are published in ISO 12640-3 and in the ICC specification.

With v4 ICC profiles, re-purposing can be accomplished by sequencing the device-to-PCS perceptual rendering intent transform of a "source" first visualization output profile with the PCS-to-device perceptual rendering intent transform of a second visualization output profile. The device-to-PCS perceptual transform from the source output profile "undoes" the previous perceptual color re-rendering from the perceptual intent reference medium to the source profile's actual output medium.

Note that use of the perceptual "undo" is appropriate only if the first visualization resulted from a perceptual rendering transformation. The rule of thumb is that the inverse of the rendering intent that was used to produce a particular visualization should be used to "undo" that visualization. Also note that even with the improved support in compliant v4 ICC profiles, subsequent visualizations can be constrained by loss of color detail in earlier transformations.

For re-purposing in general, when the destination output-referred image state gamut and viewing environment condition are "like" that of a source output-referred image state, then a colorimetric intent, with no preferential adjustment, may achieve acceptable results. (In fact, if the source and destination media are similar to the ICC perceptual intent reference medium, there should be little difference between the colorimetric and perceptual intent transforms.) On the other hand, when there are significantly different gamut constraints, and/or viewing environments, then a perceptual rendering intent, with inherent preference adjustments, can improve results. PDF/X-3 files, containing a fully populated (complete sets of PCS-to-device and device-to-PCS transforms) ICC output profile that describes the PDF output intent, support this type of re-purposing.

The goal with v4 ICC profiles is to enable blind use of perceptual intents for re-purposing. It is expected that as v4 profiling tools become more capable in generating quality perceptual color re-rendering transforms, this goal will be realized. However, in critical applications with media that are quite different from the perceptual intent reference medium, sophisticated users may find that careful, controlled application of colorimetric intents, abstract profiles, and CMS color rendering can produce better results.

13.3 Preserving an Artistic Intent through Multiple Visualizations

Preserving an artistic intent through multiple visualizations can require a combination of re-targeting and re-purposing approaches. The approach that is most likely to produce the best results in a particular situation depends on the similarities of the various actual media to each other and to the perceptual intent reference medium. When multiple independently optimized visualizations are planned in advance, alternative approaches can be considered. If a specific artistic intent is desired, particular care should be taken with the first visualization.

A large-gamut output-referred source image can be obtained by first applying the appropriate perceptual intent transform to color-render scene-referred image data to the ICC perceptual intent reference medium, and then transforming the colorimetry of that reference output-referred first visualization image to an appropriate storage color encoding such as ROMM/ ProPhoto RGB. (Note that for a color encoding to be appropriate for this use the encoding image state will match the ICC perceptual intent reference medium image state, and the profile for that color encoding will have identical perceptual and colorimetric rendering intents.) Alternatively, after using an appropriate perceptual intent transform to color-render scene-referred image data to the perceptual intent reference medium, a first "actual" visualization can be obtained by using an appropriate perceptual intent color re-rendering transform to re-render from the perceptual intent reference medium to the medium of a large-gamut output device. Using such a "superset" first visualization as the source for subsequent visualizations can improve the optimization for the subsequent visualizations, while maintaining color fidelity with the intended artistic intent.

When a color rendering to a first visualization represents a "master" image, including the artistic intent of the image creator, subsequent color transformations should not "undo" the initial perceptual intent color rendering. A subsequent actual output-referred visualization can be produced via a re-targeting approach (i.e., using colorimetric transforms) when the actual output medium is "like" the master image medium. When a subsequent actual output medium is dissimilar to the master image medium, the approach most likely to produce the best results depends on the relationships of the media to each other and to the perceptual intent reference medium. If the master image is targeted at the perceptual intent reference medium and an actual output medium is dissimilar from the perceptual intent reference medium, then the perceptual intent transform of the actual output destination profile should be used to color re-render from the perceptual intent reference medium to the actual output medium.

The case where the first actual visualization medium, the perceptual intent reference medium, and the subsequent actual output medium are all substantially different from each other is the most challenging for color management. Ideally, in this case, the device-to-PCS perceptual intent transform from the first actual visualization medium profile should be used to perform color re-rendering to the perceptual intent reference medium, and then the PCS-to-device perceptual intent transform of the subsequent actual output profile should be used to perform color re-rendering from the perceptual intent reference medium to the subsequent actual output medium. However, it is possible, perhaps likely, that the first visualization profile and next visualization profile perceptual color re-renderings may not be complementary with each other to preserve the master image artistic intent. In that case, using a specifically tuned DeviceLink profile to transform directly between the first visualization and the subsequent visualization will likely produce better results.

Table 13.1 summarizes the options for preserving artistic intent through multiple visualizations.

Note that when no related artistic intent is required among the multiple degree visualizations, then more flexibility in the final output can be obtained by retaining capture-referred (e.g., scene- or original-referred wide-gamut RGB) data to use as the source for each independent visualization color rendering or re-rendering. This enables maximum flexibility for each visualization. It should be noted that this approach can produce significantly different versions of the same image, as scene-to-picture color rendering can be quite aggressive, and involve choices such as overall lightness, contrast, tone, and saturation that go beyond the optimization of the scene to some output medium.

Table 13.1 Rendering intent and visualization options. Note that "like-ness" scale trade-offs must be evaluated for each workflow situation

First visualization	Next visualization	Rendering intent transform selection
Like the next visualization	Like the first visualization	ICC media-relative colorimetric from source profile, and from destination profile
Like the PCS reference medium	Unlike the PCS reference medium	ICC media-relative colorimetric from source profile, perceptual (designed with minimal preference adjustment) from destination profile
Unlike PCS and unlike the next visualization	Unlike the first visualization	Perceptual from source profile, perceptual from destination profile, or tuned DeviceLink

13.4 Additional Rendering Intent Sequence Examples

13.4.1 Visualization of the ICC Perceptual Intent Reference Medium Image

When it is desirable to visualize the perceptual intent reference medium rendition of a color image directly, a visualization device with capability matching or exceeding the perceptual intent reference medium is required. Given that, one can use ICC media-relative colorimetric rendering from the PCS, re-targeting the perceptual intent reference medium image to the actual output device (after correct perceptual rendering to the perceptual intent reference medium). Such visualizations should then be viewed in the reference viewing conditions (ISO 3664 condition P2) to produce the appropriate appearance.

13.4.2 Image-Specific Preferential Color Rendering

As discussed above, image-specific profiles and/or rendering intents can be used to obtain optimized preferential color renderings from a capture-referred state to the reference output-referred ICC perceptual intent reference medium. Use of image-specific color renderings should consider the need for color appearance compatibility across the various color objects intended for a particular document.

13.4.3 Color Rendering or Re-rendering from an Ambiguous Image State RGB Color Encoding

The first question when displaying color image data from an unknown image processing source is, "Has the color data been previously color rendered to an output-referred state?" The next question is, "Is the data print referred or display referred?" Certain RGB encodings inherently carry with them a particular image state: sRGB is output referred for monitor viewing; ROMM/ProPhoto RGB is output referred for the ICC perceptual reference medium print condition; Adobe RGB (1998) has historically been used to encode data relative to a variety of image states and has recently been defined as monitor display referred for future work. It can be helpful to understand the use case or workflow that produced the RGB data when inferring the color rendering image state condition. Typically, RGB data that is exchanged will have been color rendered to a first visualization and can be considered output referred. However, beyond that it may be difficult to determine whether the RGB data is optimized for print or monitor viewing. When color re-rendering from an RGB working space, both the image state of the data and the medium to which it may have been previously "color rendered" can affect the outcome of a subsequent color re-rendering. Keep in mind also that manual adjustments may have been applied to optimize the data for a particular visualization. Caution is required because repeating a scene-referred to output-referred perceptual rendering intent transformation (as described above) will degrade image quality, as will applying an inappropriate color re-rendering transformation.

A source rendering intent can be selected to be appropriate for the image data in a particular working space. For example, prior to printing typical sRGB image data, it should be re-purposed from its display-referred state to the reference print output-referred image state corresponding to the ICC perceptual intent reference medium. On the other hand, if a user has edited Adobe RGB image data to produce a desired appearance on a print medium, a relative colorimetric source rendering intent may be appropriate when transforming for print.

When selecting the "next visualization" destination rendering intent for a previously color-rendered (output-referred) RGB encoded image, as above, color re-rendering from the perceptual intent reference medium to an actual output visualization encoding can be media relative, or absolute colorimetric when the actual output visualization gamut extent and tone range are similar to the reference medium gamut extent and tone range. When the actual output visualization gamut extent and tone range are significantly different from those for the reference medium, then perceptual rendering may provide an improved result.

13.4.4 Color Re-rendering of Computer-Generated Imagery

Use of the perceptual rendering intent in reproducing computer-generated color infers the computer display as the "original" capture device. The computer display "synthetic original" (original-referred image state) can be preferentially color re-rendered to the ICC perceptual intent reference medium using the perceptual rendering intent of a v4-compliant input profile for the computer display. Consideration of the rendering intent to use from the perceptual intent reference medium to the "next visualization" actual output encoding is similar to that discussed above.

References

[1] ISO 22028-1:2004. *Photography and graphic technology – Extended colour encodings for digital image storage, manipulation and interchange – Part 1: Architecture and requirements.* International Organization for Standardization, Geneva

Part Three

Workflows

14

Using ICC Profiles with Digital Camera Images

There are two kinds of ICC profiles that can apply to image files created by digital cameras: color space profiles and input profiles.

It is important to understand that, except for applications like copying art and product photography where the picture is supposed to exactly match the original captured, pictures usually do not match the scene from a color measurement, or even necessarily from an appearance standpoint. Typically, the contrast and color saturation will be boosted (especially in the mid-tones) to the extent allowed by the reproduction medium (and consistent with a "natural" appearance in the expected viewing conditions), and specular highlights will be compressed for printing and viewing on typical displays. This scene-to-picture color processing is called "color rendering" (as defined in the Glossary in Chapter 8 and in ISO 22028-1). More complicated adjustments are also performed, especially in cameras aimed at the consumer market. For example, some cameras individually color render each scene, considering its dynamic range and key. "Digital scene re-lighting" algorithms that attempt to compensate for uneven scene illumination are also used.

When a camera is producing image files that are based on standard color encodings, such as sRGB, Adobe RGB (1998), or ProPhoto RGB (also known as ROMM RGB), the color rendering is being performed by the camera. The encoded image does not represent the original scene, but rather the camera's attempt to create and encode a pleasing reproduction of the scene (i.e., a picture). These encodings are called "standard output referred" since they encode the colorimetry of the picture on a standard output reference medium. In the case of sRGB, the reference medium is a standard CRT display. In the case of ProPhoto (ROMM) RGB, the reference medium is the same as the ICC perceptual intent reference medium reflection print. The Adobe RGB reference display is a 160 cd/m^2 additive display with a D65 white point and a large color gamut based on the Adobe RGB (1998) primaries, viewed in a dim surround, with the same luminance ratio as the ICC perceptual intent reference medium. Other details of this reference display can be found in the Adobe RGB (1998) specification published by Adobe Systems.

Color Management: Understanding and Using ICC Profiles   Edited by Phil Green
© 2010 John Wiley & Sons, Ltd

So, when a digital camera creates an image file using a standard output color encoding, the correct ICC profile to associate with that file is the profile for the color encoding used, not a profile for the camera itself. If one tries to create camera profiles for such files by photographing a target, the results will generally be suboptimal because the profile will in effect be trying to undo the color rendering applied by the camera to get back to the scene. There will almost always be errors, in part because of the limitations of reflection target-based characterization. Also, for most applications the actual scene color will often be less pleasing than the color-rendered picture. Furthermore, all cameras apply white balancing, so a different profile is required for each white balance setting. The placement of the characterization target white on the tone scale can also produce different results. Cameras that apply digital scene re-lighting will have characteristics that vary across the image, and therefore cannot be undone using an ICC profile which is applied to the image as a whole. Finally, for cameras that perform image-specific color rendering, the profile created is only certain to be correct for the image of the target, since the color rendering applied may be different for other scenes photographed.

There is also the case where a camera generates files containing raw or scene-referred image data. If the raw image data results from capture using a color filter array (e.g., the red, green, and blue color values are captured by a sensor array in which the individual photosites have red, green, or blue filters), a camera raw processing application is needed to create a viewable color image. In most cases, these applications (e.g., Adobe Photoshop camera raw) create standard output-referred images, as would the camera, though not necessarily with the same color rendering. Camera raw processing is valuable because the user can guide the color processing applied to the raw image data, thereby eliminating the losses that result from incorrect white balancing or color rendering. These choices can be made without loss, after the picture is taken, to create the finished image file.

A very few cameras and camera raw processing applications generate scene- (or focal plane)-referred image data. The cameras are typically professional camera backs that are designed for studio use. In this case it can be appropriate to create a camera profile that represents the scene in the ICC PCS using the colorimetric rendering intents. A simple reflection target-based characterization will often not produce the best results, and it may be better to use the camera's spectral sensitivities to calculate the transformation matrix, which will typically depend on the white balance. Ideally, this calculation will be optimized to the spectral properties expected for the scene to be photographed. The perceptual intent of these true camera profiles should include color rendering to the ICC perceptual intent reference medium, and can be used for general photography. Camera profiles will typically be specific to particular shooting conditions (illumination, camera exposure settings, scene dynamic range, key, etc.).

In summary, in most cases the profiles that should be used with digital camera images are the appropriate standard color space profiles. It is only when professional cameras that produce scene-referred image data are used that true camera profiles are appropriate. Reproducing relative scene colorimetry or appearance is primarily appropriate for specialized applications such as copy work, and product or catalogue photography where scene color matching is the reproduction goal. Expressing relative scene colorimetry or appearance may also be appropriate in applications where the primary color rendering will be applied manually or with special purpose tools later in the reproduction process.

The camera color rendering that is applied is sometimes inadequate to meet user needs. Camera profiles provide a way to apply color transformations, and in some cases there are controls in the profile creation software that allow photographers to create custom-modified profiles to accomplish a specific purpose. ICC profiles can be used in this way to correct for color rendering deficiencies in specific images or groups of images. However, using camera profiles to compensate for inadequate color rendering can cause problems in profile management, workflow, and interoperability, and can also contribute to user dissatisfaction. It is also somewhat misleading to think of these profiles as camera profiles, because in most cases they are essentially image correction profiles, or color re-rendering profiles.

ICC color management workflows by default assume that the colorimetry expressed in the PCS is of a picture that has already been color rendered to an output medium, and not of an original scene. The CIIS tag makes it possible to indicate that the colorimetry represented in the PCS by a colorimetric intent transform is scene-referred colorimetry. Where applications make the default assumption that color rendering has already been performed, scene-referred colorimetry may not produce preferred results. Applications used in workflows that include scene-referred images should be able to interpret the CIIS tag and either use the perceptual rendering intent or enable appropriate color rendering of the image. This is especially important in the reproduction of highlights: many scenes contain highlights that are brighter than the tone in the scene that is reproduced as white in a picture, and the color rendering process should be able to select the tone in the scene that is considered "edge of white," and apply graceful compression of brighter tones to fit on the reproduction medium (between the "edge of white" tone and the medium white).

The ICC Digital Photography Working Group is addressing the use of ICC profiles in digital photography applications, and has made considerable progress in demonstrating the use of scene-referred images in color management workflows. Narrow-band emissive targets, characterization targets, and profiling tools are available from some sources, although colorimetric intents will still be illumination specific, and perceptual intents will optimally be scene specific. Some would argue that scene-to-picture color rendering should be restricted to in-camera processing and camera raw processing applications, and correction of color rendering deficiencies limited to image editing applications.

Creating an Input Profile for a Digital Camera
Scenes photographed by a digital camera have variable illumination, and a camera profile will therefore be scene specific. For this reason the usual practice is to convert the image data to either a standard output-referred color space encoding such as sRGB or ROMM RGB, or to a standard input-referred color space encoding such as RIMM RGB. The profile for the standard color space is then used in preference to a scene-specific camera profile. An example of a standard color space encoding profile is shown below.

Tag	Size (bytes)	Value
"desc"	84	ISO 22028-2 ROMM RGB profile
"A2B0"	212	v4 lutAToBType with M curves, 3×4 matrix, and B curves
"B2A0"	212	v4 lutBToAType with M curves, 3×4 matrix, and B curves
"wtpt"	20	[0.858 09, 0.89, 0.734 21]
"cprt"	88	Copyright 2006 Hewlett Packard
"chad"	44	Identity matrix

If the image data is in a scene-referred image state, it can be processed as representing the colorimetry or appearance of the actual scene. Because such image data will not have undergone rendering to a standard color encoding or an output-referred image state, it will normally require further processing to generate a pleasing rendering of the scene. The image state should be flagged in the profile by the use of the colorimetricIntentImageStateTag (signature "ciis"), which currently has signature values as follows:

scene colorimetry estimates: "scoe"
scene appearance estimates: "sape"
focal plane colorimetry estimates: "fpce"
reflection hard copy original colorimetry: "rhoc"
reflection print output colorimetry: "rpoc"

In many cases the camera will record a scene maximum white that has a considerably higher luminance than that of a perfect diffuser under the same viewing conditions. To accommodate this, and thus preserve high dynamic range data for later processing, the ICC specification recommends that for scene-referred data the mediaWhitePointTag Y value is relative to the Y value of the scene adopted white, and can be as high as 2.0. The scene dynamic range can be recorded in the profile through the use of the viewingConditionsType tag. An example of a scene-referred camera profile is illustrated below. This includes both colorimetric and perceptual intents in the device encoding-to-PCS direction.

Tag	Size (bytes)	Value
"desc"	114	Nikon D70 camera RGB – fluorescent WB – HR2
"A2B0"	24 916	v4 lutAToBType with A curves, 3D CLUT, M curves, 3 × 4 matrix, and B curves
"A2B1"	248	v4 lutAToBType with M curves, 3 × 4 matrix, and B curves
"vued"	136	"Daylight adopted white average surround"
"view"	36	XYZ of illuminant (cd/m^2): [19 136, 20 000, 18 428] XYZ of surround (cd/m^2): [3444, 3600, 3317] Illuminant type: D55
"wtpt"	20	[1.928 41, 2.0, 1.649 80]
"cprt"	36	"none"
"chad"	44	Identity matrix
"ciis"	12	"scoe"

A standard input-referred profile with a media white point higher than that of the perfect diffuser is linear_RIMM-RGB_v4.icc, which can be downloaded from the ICC web site. This includes both colorimetric and perceptual intents in the device encoding-to-PCS direction and a BToA1 tag to invert PCS data back to the device encoding.

Tag	Size (bytes)	Value
"desc"	80	Linear RIMM RGB profile v4
"A2B0"	24 832	v4 lutAToBType with A curves, 3D CLUT, M curves, 3 × 4 matrix, and B curves
"A2B1"	152	v4 lutAToBType with M curves, 3 × 4 matrix, and B curves
"B2A1"	152	v4 lutBToAType with M curves, 3 × 4 matrix, and B curves
"wtpt"	20	[1.92 841, 2.0, 1.649 80]
"cprt"	88	Copyright Hewlett Packard 2007
"chad"	44	Identity matrix

15

RGB Color-Managed Workflow Example

15.1 Overview

RGB source image data such as photos, illustrations, and digital art are routinely "re-purposed" (redirected for different outputs). The workflow that best supports such re-purposing is to start with a master image which can then be used with many different output processes. Standard RGB color encodings are well suited to these requirements, as they are less tied to a specific output device and are thus more device independent.

In the following workflow example, a master RGB image is color corrected and archived in RGB form. From there, it may end up in various output forms such as a billboard, a piece of art for a web site, or a newspaper advertisement. Regardless of the output medium, a single RGB file can be the source in the production of pleasing results for each medium.

To retain the ability to re-purpose image content, it is essential to preserve the source encoding until as late as possible in the workflow, using ICC profiles to implement the necessary conversions. This workflow is often referred to as "late binding."

An example RGB workflow is illustrated in Figure 15.1, and the purpose of this chapter is to provide outline recommendations on implementing a late binding RGB workflow. Information on making profiles, and on tools for this purpose, can also be found on the ICC web site at http://www.color.org/creatingprofiles.html.

15.2 Prerequisites

To achieve an accurate and efficient RGB workflow, each color device in the workflow must be calibrated. This is the process by which a device is returned to known conditions, or deviations from these conditions are compensated for. It is followed by characterization, which is the process of sampling the device encoding in order to generate characterization data and a model of the relationship between the particular device and CIE colorimetry. In a color management system (CMS) or workflow, a specific profile is based on the characterization data but in

Color Management: Understanding and Using ICC Profiles   Edited by Phil Green
© 2010 John Wiley & Sons, Ltd

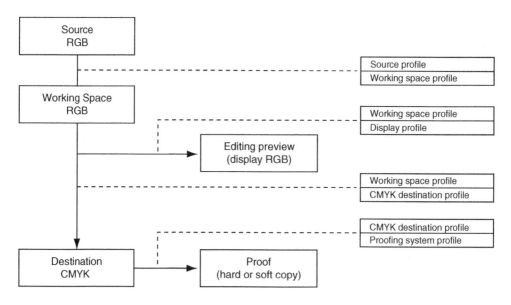

Figure 15.1   RGB workflow example

addition provides additional manipulation of the data for specific needs. For example, a PCS to device CMYK output profile includes gamut compression, tone scale adjustment, black generation, and so on. The process of calibrating and characterizing may be slightly different for each type of device.

A standard color management module (CMM), sometimes called a color engine, should be chosen for the workflow. A CMM is the software component that transforms the color values in the source color encoding into color values for the destination color encoding using ICC profiles. CMMs are provided, for example, by Apple in Mac OS, by Microsoft in Windows, and by vendors such as Adobe and Heidelberg.

15.3 Source Profiling

Source profiles for workflow input devices such as scanners and digital cameras should be located, or created, as needed. In the great majority of cases, a scanner or digital camera will create RGB images in a standard output-referred color encoding, such as sRGB, Adobe RGB (1998), or ROMM (ProPhoto) RGB. In this case a profile for the color encoding used should be assigned as the source profile and embedded in the image.

In some cases the source color encoding will be a raw file. Scanners that produce raw data should be profiled using an appropriate characterization target, as specified in ISO 12641. Raw digital camera files should be processed to some standard color encoding using a camera raw converter. This processing includes the white balancing and color rendering, and is where most of the editorial choices are made.

In most workflows source data is output referred. Some source data is scene referred, using color encodings such as scRGB and RIMM RGB. In such images white balancing has been performed, but not color rendering, and the source data is defined with respect to the colorimetry of the original scene. If a raw camera RGB or scene-referred file contains an embedded ICC v4 profile, it should be used to convert to a standard output-referred color encoding before proceeding. Generally the perceptual rendering intent should be used.

The source RGB color encoding should be chosen based on the anticipated destination media gamuts; sRGB is well suited to typical display gamuts and is therefore the best choice for web content. It may also be a good choice for lower end printed material. Adobe RGB (1998) is a common choice for content that will be presented on wide-gamut displays and for mainstream printing; sRGB and Adobe RGB (1998) are also used as working spaces for image editing. ROMM RGB is the best choice for high-end printing on large-gamut media, but only experts should use it as a working space as some code values are not allowed, and 16 bits per channel are recommended. RGB images should always contain an ICC profile describing the current color encoding that the image is rendered to.

15.4 Display Profiles

Natively, different monitors often display color differently and may not fully encompass the gamut of CMYK output. To compensate, the signals sent to a monitor must be color managed using an ICC profile, either one provided by the manufacturer or one created using display profile creation tools.

15.5 Profiles for Proofing and Printing

Print destination profiles are available for standard printing conditions (defined in ISO 12647-2 and ISO 12647-3) on the ICC Profile Registry. These profiles are recommended for use when the actual printing condition corresponds to one of the conditions for which these profiles were generated. Printer profiles may also be provided by the manufacturers of digital printers, such as inkjet printers and digital presses.

Where such a profile is not available, users may wish to generate a custom profile. Where the printing condition corresponds to one of the printing conditions in the ICC Characterization Data Registry, this data should be used in preference to preparing custom characterization data.

Where published characterization data does not exist for the target printing condition, users will first need to obtain such data by measurement of printed targets before generating the profile. The printing system should be linearized prior to characterization. CMYK output devices can be characterized using a target such as ECI 2002 or ISO 12642-2 (also known as IT8.7/4). Randomly arranged (as opposed to visually arranged) targets work best on most press types. Print two or more random targets, rotated at 90° to each other, close to the center of the page and not in-line with each other. Measure and average the results to provide the most accurate characterization data.

When generating a profile in an ICC profiling application, the appropriate amount of black generation and ink limiting should be used. In lithographic printing ink limiting may vary between 200% and 350% total area coverage (TAC), depending on the printing process and paper, In ink jet printing the TAC will typically need be at the lower end of this range unless ink limiting has been applied independently in the RIP. Default black generation settings for printing processes are provided by profiling applications, and recommendations on such settings are available from industry bodies such as SWOP, fogra, ECI, SNAP, and ifra.

In proofing, the aim is to simulate the output of the final reproduction process. The profile for the proofing system is generated from targets printed on the proofing stock, using the methods described above. The image will usually be converted to the print destination profile, using a perceptual or colorimetric rendering intent according to the goals of the reproduction, and then converted to the proofing system profile using a colorimetric intent.

15.6 Assigning Source Profiles

In a color-managed workflow it is very important that each image has an associated profile. Documents containing embedded profiles are referred to as tagged files (e.g., scanned art tagged with the scanner's profile, or digital camera files tagged with the sRGB or Adobe RGB (1998) profiles). Image files with the proper ICC profile embedded provide a computer's color management system with valuable information about how individual image and graphic objects should be displayed, processed, and printed.

15.7 Conversion to an RGB Color Encoding

Prior to image editing, convert to an RGB color encoding or working space, as shown in Figure 15.1. It is not wise to edit colors in an output device-specific color space. Choose a working space based on the considerations noted above.

15.8 Soft Proofing

Using ICC profiles allows you to soft-proof the printed results on your monitor. In Adobe applications, choose "Custom Proof Setup" from the View menu. Select the proper destination profile in the Device to Simulate menu. Use "Black Point Compensation" and choose either "Perceptual" or "Relative Colorimetric" as the "Intent." Check "Simulate Black Ink and Paper Color" to preview shadow details and the overall effect of printing on the chosen stock. Soft proofing in this manner provides a reasonable viewing environment for image evaluation and editing. This setup does not necessarily provide an accurate appearance match between the monitor and the equivalent hard copy output.

15.9 Adjusting Color

Color management tries to maintain the artistic intent of the original image, with the degree of appearance matching depending on the rendering intent selected. The ICC-absolute colorimetric rendering intent attempts to produce an appearance match within the gamut of the destination medium, clipping any out-of-gamut colors. The media-relative colorimetric rendering intent maps the colors relative to the source medium white to the destination medium white, avoiding both highlight clipping and muddy highlights. Black point compensation similarly scales the black point and shadows. Perceptual rendering intents aim for pleasing reproductions while preserving the original artistic intent and taking the capabilities of the printer into account, but in doing so may produce even larger differences in appearance. In v2 workflows some profile perceptual transforms result in poor results due to difference reference gamut assumptions. Further color adjustment may also be needed in order to achieve the desired appearance and/or creative goals for individual images.

Images that require extensive editing should be acquired and archived at higher bit depth, 16 bits rather than 8 bits per channel, whenever possible. This minimizes the possibility of quantization errors (contouring or banding) being introduced during the editing or conversion processes.

Images should be acquired as close as possible to the size that they are to be used, but if they require further resizing, it should be done at this time. Any edits in CIELAB color space should be done at 16 bits per channel to maintain precision. After editing is complete, the image can be converted to 8 bits per channel for archiving.

When editing images, the destination profile that represents the current destination device should be selected as the CMYK working space as this will allow the results of editing to be previewed.

15.10 Conversion to Destination Color Space

Images whose colors all lie within the gamut of the destination profile can easily be converted using the media-relative colorimetric rendering intent with black point compensation. Images that have colors that are outside of the gamut of the destination profile may benefit from a color re-rendering, so the perceptual rendering intent may be selected to prevent simply clipping the out-of-gamut colors. Out-of-gamut colors can be previewed in Adobe Photoshop by making sure that the destination profile is selected in the Color Settings control dialog, and then selecting "Gamut Warning" from the View menu. This will highlight any areas in the image that will be clipped when converting to the destination color space.

Even if all the source colors are inside the destination gamut, a user might prefer the reproduction resulting from a perceptual rendering intent over a colorimetric reproduction as the perceptual re-rendering may produce a more pleasing result.

Color server software can also be used to convert pages on the fly. This is a highly efficient workflow, which works best if images are converted into a common RGB color encoding and a well-tested profile and rendering intent are used for the final conversion to CMYK.

15.11 Sharpening

Sharpening should be process specific and is ideally performed after the image has been encoded in the final device space. Unsharp masking will enhance edge detail, and should if possible be applied to the lightness channel or in "luminosity" mode. Sharpening parameters will vary according to the output media, and will be different for billboards, for example, than for the Internet or newsprint. In an RGB workflow, the specific software solution will dictate whether sharpening can be done in CMYK immediately prior to proofing, or must be done earlier in the RGB stages of the workflow. Late binding sharpening is the most desirable solution and allows greater workflow flexibility.

15.12 Additional Editing

Black-only drop shadow creation can be done at this time. Additional spot colors or high GCR areas should be defined and composited into the image file.

Legacy CMYK files can be converted to RGB with very little loss in color fidelity, by using ICC profiles with media-relative colorimetric rendering intent and black point compensation. Convert the data to the workflow's device-independent color space (typically the current RGB working space). These files can now be re-purposed just like the rest of the files in the RGB workflow.

Page files can be built from RGB archives and CMYK images, created as a first-generation conversion from the RGB parent files.

16

Issues in CMYK Workflows

16.1 Introduction

In the graphic arts workflow, CMYK is a common choice of ink set for images and graphics. While there are other possible color sets, especially in the packaging market, CMYK is by far the predominant set used across most of the graphic arts industry for final production and printing. While some of the points in this chapter are universally applicable to all inking systems, the chapter deals primarily with the special issues inherent in a CMYK workflow and the choices which a user faces when processing CMYK in a color-managed workflow. Other issues in graphic arts color reproduction are considered in Chapter 18.

16.2 Background and Relevance of Black

The CMYK system uses three chromatic primaries (cyan, magenta, and yellow) plus an achromat (black) in different combinations to make up all colors. While it is possible to use just CMY, this is usually not desirable in real printing systems for a number of reasons:

- The spectral properties of the individual CMY colorants may not combine to give the desired absorbence, resulting in a muddy brown instead of black.
- Black is less susceptible to press or proofer variations affecting color balance when printing neutral grays.
- In many printing systems it is possible to produce a darker gray or black using black ink rather than some combination of CMY alone.
- Where the gray component of a color can be replaced with black, there can be significant ink savings.

Black plays a large part in the perceived contrast within images. The black colorant is significantly less expensive than CMY, providing an economic incentive to use black in concert with or in place of CMY. In addition, type is usually printed in black and the use of a single colorant to represent black instead of three (CMY) avoids trap and register problems, and also avoids possible half-toning of content that would occur with CMY.

Color Management: Understanding and Using ICC Profiles   Edited by Phil Green
© 2010 John Wiley & Sons, Ltd

Because of the role of black in image contrast, color separation used to be something of a "black art" requiring a high degree of operator skill, both in determining the amount of black as well as the physical process (film or electronic) used to produce the color separation. The effort invested in producing the color separation strongly discouraged most people from changing the CMYK separations once they had been produced. In fact, many print buyers would insist on color separation integrity, hence the rise of half-tone proofing. In the traditional CMYK workflow, color separation integrity typically came at the expense of color fidelity.

Working with CMYK files in a color-managed workflow where color transformations of content can take place in multiple locations, either for re-separation or, more commonly, for proofing, can be problematic. The classic problem for such workflows is the accidental introduction of an unintended CMYK–CMYK color transformation where one was not desired, destroying the structure of the CMYK file in the process. Common problems which are observed include:

- Pure black (0–0–0–K) turns into four-color C–M–Y–K color build, with resulting color shift, misregister, and/or trap implications. Also known as the black type problem.
- Black channel proportionality relative to CMY changes, resulting in a change in apparent contrast or TAC. Also known as the shape problem.
- Unintentional color management of CMYK which happens silently and untraceably in a workflow with several hand-offs of data.

The problems with the color-managed CMYK workflow arise because of both color transformation mathematics as well as workflow data handling issues. Workflow data handling issues revolve around the choice of when in the workflow one chooses to perform a particular color transformation.

Workflow data handling is partially addressed by the PDF/X-1 data interchange standard (ISO 15930-1:2001(E)) provided the OutputIntents array information is set correctly. PDF/X-1a ensures that only print-ready, color-separated CMYK exists within a given graphic data file. The OutputIntents array in a PDF/X-1 file specifies the characterized printing condition that the CMYK data in the PDF/X-1 file is destined for.

It is intended that in a PDF/X-1a workflow, the CMYK content should remain basically untouched once it has been created. The OutputIntents array then specifies the source profile used any time an ICC color transformation from CMYK is required (e.g., for proofing). This implied usage of the PDF/X-1 file, where the CMYK is intended to serve as a digital master, will tend to keep data intact provided that all handling applications respect the PDF/X-1a specification.

16.2.1 CMYK Conversion Styles

There are a number of different methods of performing CMYK → CMYK conversion. While this section does not provide full implementation details, it covers most of the main methods used in the industry along with some commentary on the rationale and issues for each method.

16.2.2 Do Nothing to CMYK

This is the traditional CMYK handling technique. In order to effect color changes for proofing and presswork, the only modification allowed is a dot gain modification by means of applying a transfer curve.

The transfer curve method performs a one-dimensional transformation on CMYK: cyan is multiplied by an amount which varies depending on the cyan tint only, magenta is multiplied by an amount which varies depending on the magenta tint only, and so on.

The chief disadvantages of transfer curves are that they are limited in the extent to which color can be modified (sometimes it is necessary to introduce other colorants, especially for proofers) and they are an unnecessarily crude tool to use if one is trying to perform fine color manipulation.

The chief advantage of maniulating color by means of a transfer curve is that it tends to leave the relationships between the CMYK colorants intact. For example, if a certain color has no cyan in it, no cyan will be introduced by the transfer curves. Moreover, transfer curves are traditionally only applied at the final output stage, whether one is making proofs or plates. (In the cases of film exposure and press printing, the processes themselves introduce some dot gain.)

In the workflow where nothing is done to the files except on output, there is fairly little risk of inadvertently modifying the CMYK separations.

16.2.3 CMYK → PCS → CMYK

This is the traditional ICC-based CMYK handling technique. This transformation is composed of two parts. The CMYK is first transformed into a three-dimensional colorimetric profile connection space (PCS), either CIE $L^*a^*b^*$ or XYZ, from the source CMYK color space, and then transformed back from PCS into the destination CMYK color space.

While this method will usually result in the colorimetry being correct, with no inherent restriction on how close colors can be matched, other than the relative sizes of the source and destination gamuts, there is a fairly substantial problem for the CMYK workflow.

The problem with this method is that there is no unique solution to the PCS → CMYK mapping. While the source-to-PCS part of the transform (CMYK → PCS) is unique since there will only be one PCS value for each CMYK combination, there are many possible choices for the PCS-to-destination step (PCS → CMYK).

The original choices for black separation are destroyed in this process. For example, a single color black (0–0–0–K) on the source side turns into a four-color CMYK build on the destination side, with implications for both color and trap.

The workflow which involves an explicit transformation through colorimetric space is the source of many of the problems encountered with CMYK data in a color-managed workflow environment.

16.2.4 Preserve Pure Black

Perhaps the simplest of the solutions, this approach involves preserving pure black (0–0–0–K) as pure black, while performing regular ICC color management on everything else. Depending on the system, the output black tint might be further adjusted in order to obtain the correct lightness (L^*) value at the destination device.

While this approach resolves the black type problem, it does not address the shape problem in images. It can also introduce artifacts, as there will tend to be contouring in blends between pure black and other colors.

A special case of this, which a couple of vendors have implemented in the past, is where only 0–0–0–100 is left alone as 0–0–0–100, on the assumption that black text will always be 100% black and should always be represented with 100% black on the destination device, regardless of the actual destination color gamut.

This workflow is easy to implement but suffers from some drawbacks, since it only solves a part of the problem while at the same time introducing new problems.

16.2.5 CMYK → CMYK Workflow

A more sophisticated approach will not only preserve pure black solids and tints as pure black, but also attempt to preserve the black proportionality relative to CMY across the rest of the color space as well, while still performing ICC-based color management.

One way of looking at this is that it is essentially the same as the CMYK → PCS → CMYK approach, except that one chooses PCS → CMYK where the destination CMYK is as close as possible to the source CMYK, while still being colorimetrically accurate.

There are several implementations of this type of workflow from multiple vendors, all of which approach this problem slightly differently in the details. Most use ICC DeviceLink profiles, while some vendors have a solution built into their CMM. In some workflows, it is possible to impose additional constraints such as preserving pure colors (C, M, Y individually) in addition to preserving pure black. Some colorimetric accuracy is taken away for each additional constraint. Preserving pure colors is important if one wishes to avoid things like scum dots in solid yellow, where they will be highly visible.

Because the black proportionality is more or less maintained, there is less damage to the color separations with successive conversions. Hence this workflow is probably the best solution to the problem of maintaining color separation integrity in a color-managed workflow. The downside is that variability in allowed constraints as well as a multitude of vendor–proprietary solutions reduce the interoperability of these workflows across vendors relative to the CMYK → PCS → CMYK workflow.

16.3 Summary

There is no simple answer to the color-managed CMYK workflow issue even though there are several technologies available on the market which address the problem. The most difficult aspect of the problem that users face is deciding upon and enforcing particular color handling choices at different points in the workflow. Once this has been considered, there are several CMYK → CMYK solutions on the market which can help users implement a robust color-managed CMYK workflow.

17

Orchestrating Color – Tools and Capabilities

The process primaries CMYK continue to be the basis of most data exchange for the graphic arts. However, color management is being increasingly used in the creation of CMYK data, and even when color management is not used in this way, it is being used to identify the printing conditions for which the CMYK data was intended. The PDF/X-1a file format, ISO 15930-1 [1], requires pointers to standard characterization data to be included as part of the file. The preferred registry that is identified in the PDF/X standards is the ICC Characterization Data Registry at http://www.color.org/chardata. Where the expected printing does not match a registered characterized printing condition, a destination profile must be included.

This has placed much more emphasis on the ICC Characterization Data Registry and the characterized printing condition data that is identified in that registry. In the registry, we have a single location to point to where established sets of data that relate CMYK input values to printed color are identified. At the same time the main graphic arts data exchange formats now require such information.

17.1 Exchange of Color-Managed Data

There is increasing interest in exchanging three-component data, more so in Europe and newspaper applications. There are of course many different RGB encodings from which to choose, and color management is required to provide a basis for data exchange between such encodings and the final CMYK.

In September 2002, the first graphic arts data exchange standard that fully enabled the exchange of color-managed data came into existence. PDF/X-3 (ISO 15930-3:2002 [2] which was updated by ISO 15930-6:2003 [3]) represents a major step forward and allows the exchange of fully defined three-component data for graphic arts applications. It requires the use of ICC destination profiles to identify the intended output condition and to define the data conversion between the ICC profile connection space (PCS) and the input code values of the intended printing device. It also makes provision for source profiles to be used to define the

Color Management: Understanding and Using ICC Profiles   Edited by Phil Green
© 2010 John Wiley & Sons, Ltd

specific three-component data (RGB) being exchanged. However, the standard does not say what three-component data should be used, nor does it recommend profiles. These are all user choices.

The same application areas that are encouraging the exchange of three-component color-managed data are also increasingly accepting soft proofing on the color monitor. Some of the issues (and potential pitfalls if not handled properly) involved in exchanging three-component color-managed data based on soft proofing are considered below.

17.1.1 Display

The display must be well controlled and calibrated and must have a profile that will compress or clip data so that it will fit the gamut of the display device. The typical monitor is approximately sRGB, although larger gamut displays are increasingly being used in editing and soft proofing. Intermediate working spaces are usually a large-gamut RGB such as Adobe RGB (1998). This is important since the gamut of typical CMYK printing exceeds the gamut of sRGB in some parts of color space. Appearance modeling must also be used to make a relatively dim self-luminous display look like a reflection print viewed under high illumination.

17.1.2 Profile Interchangeability

Although the format for ICC profiles is defined in the ICC profile specification, the transforms included in source and destination profile perceptual rendering intents are based on proprietary technology. Profiles from one vendor will not produce the same results as those from another vendor, nor should they be expected to. Some of those differences are what allow vendors to differentiate themselves. Different destination profile perceptual CMYK rendering intents, even from the same vendor, may handle tone reproduction, gamut compression, and black generation significantly differently. That is why PDF/X-3 says that the profile included as part of the data exchange should be used to render the data to CMYK.

Even with colorimetric profiles, different colorimetric profiles should produce colorimetric values that are close to each other, but they all handle colors near the gamut limit differently. In addition, in going from PCS to CMYK data, each vendor has unique color separation and black generation algorithms – the color should be close, but the components will be different.

17.1.3 Image Assembly

The issue of the assembly of multiple files using three-component color-managed data has not yet been cleanly solved by the standards community or by the application vendors. We can associate a source profile or color space definition with each object. However, we cannot associate any other type of profile with individual objects. There can be only one destination profile for any single PDF file. This applies to all objects within the file.

If we want to treat images differently within the same file, for example, high-key vs. low-key tone reproduction in a destination profile, we cannot do that with output profiles. Such adjustments must be accomplished with source profiles or in the editing of the original file.

Further, if multiple files are prepared for the same characterized printing condition but use different output profiles (or profiles from different vendors), they cannot be combined without additional processing. The caution in the PDF/X-3 application notes says:

If device-independent color data is used in PDF/X-3 files, the profile included in the OutputIntent of each file must be compared to those in all other files to be assembled together. Where all profiles are identical, the files may be assembled directly, retaining device independent colors. If different profiles are used, then colors must be transformed to the output device color space prior to assembly to ensure that the correct gamut and tone compression is performed for each entity.

17.1.4 Black Channel Preservation

To convert CMYK data from one device to another (where the gamuts are the same or close to each other), combining a colorimetric device-to-PCS transform for the first device with the colorimetric PCS-to-device transform for the second device should yield the correct colorimetric results. And it does, except that the color separation scheme and black printer will be that which were incorporated in the profile for the second device and may not bear any relationship to the initial CMYK. If this is for a non-half-tone proofing device, it is probably acceptable, but if the black-to-color relationship is important, then some other transform is required. A number of applications have the ability to create black-preserving device link transforms, as discussed in the previous chapter.

This is the classic problem that is faced by proofing systems and those systems that want to optimize CMYK data for a specific output device. Here the gamuts are correctly maintained by process control of solid ink density, but differences in tone-value increase, trapping, and so on result in different CMYK input being required for within-gamut colors. Using the gravure process to match offset SWOP data is an example of this situation.

17.1.5 Re-purposing and Re-targeting

Re-purposing occurs when output is sent to a device with a different gamut than the gamut it was initially prepared for, for example, CMYK publication data to a web display. Re-targeting is sending data to a device with the same gamut but a different encoding. Re-purposing and re-targeting are discussed further in Chapter 4.

In re-purposing, the first decision that must be made is whether the appearance in the initial output mode (e.g., CMYK publication) should be preserved. If so, the output data must be colorimetrically converted back to PCS and then either a colorimetric or perceptual output profile used to convert to the new destination, depending on the relative size of the color gamuts of the initial and new destinations. If the appearance in the initial output is not significant, then a new destination profile can be substituted, but the image should probably be reproofed for the new output condition to be sure the intent of the designer is preserved in the new output color space.

17.2 PDF/X

In PDF/X-4 [4] and PDF/X-5 [5], the major changes to the previous PDF/X version were focused on giving additional flexibility to data exchange. They both added the ability to provide an external reference to profiles rather then embed them in the PDF/X file itself. In addition, PDF/X-5 included the ability to externally reference output intent ICC profiles for *n*-colorant print characterizations.

17.3 Characterization Data and Reference Printing Conditions

The term "characterization data," as used here, simply describes the relationship between input CMYK tone values and the color on the printed sheet when printed according to a given printing definition. Thus, a specific set of characterization data is tied to a specific printing definition. Most characterization data uses either the 928 patch IT8.7/3 target or the 1617 patch IT8.7/4 target, and represents a robust description of its associated printing definition.

Initially, characterization data was based on careful test printing in accordance with the printing definition being characterized. More recently, data manipulation and data smoothing have been used to take characterization data created for one printing process definition and modify it so that it matches a more pleasing set of aims, or even a different set of process aims entirely.

Current data manipulation software allows a great deal of flexibility in adjusting data to match predefined aims for solids, two-color overprints, and tone-value increase curves. Because so many of the variables of real printing (one- and two-color trapping, ink transparency, etc.) are poorly defined, created characterization data needs to be either based on or evaluated through practical testing.

17.4 How Is Characterization Data Used?

In today's world of color management, digital proofing, digital plate making, and even digital printing, a set of characterization data associated with a particular printing definition has become the definition of that printing condition. Because each industry group worldwide wishes to fine-tune the generic printing definitions of ISO 12647 [6] to their own interpretation, we currently have several sets of characterization data that are all aimed at essentially the same set of conditions in ISO 12647 but vary slightly with respect to each other.

Nevertheless, characterization data has become the communication interface between design/preparation, proofing, and printing. Nothing emphasizes this more than the title of the recently approved ISO 10128 [7]: *Methods of Adjustment of the Color Reproduction of a Printing System to Match a Set of Characterization Data.*

Many different color management profiles can be created from any set of characterization data. Organizations creating characterization data are also preparing and approving ICC profiles made with this data, often in the form of a single profile used as the primary reference. It is important to recognize that any given profile severely restricts the characterization data upon which it is based. Any single CMYK output profile contains a specific methodology for and a single level of GCR, one total-dot-area setting, one color separation methodology, one method of gamut compression, one tone reproduction curve, and so on. Different profiles can contain different combinations of these parameters and thus provide multiple options to adapt input data to a particular set of characterization data.

Today, virtually all content data printed is transferred between preparation and printing as electronic data. Further, computational tools exist to manipulate that data using either single channel manipulations (the matching tone-value curves or use of near-neutral scales of ISO 10128) or in multi-dimensional transforms using color management as defined in the ICC specification. These tools are primarily focused on maintaining the appearance of within-gamut colors by adjusting the values of the overprint colors.

The one aspect of this data that cannot be predictably manipulated, although color management can do a reasonable job, is color gamut. The outer gamut of the printable color volume is primarily defined by the combination of the color of the paper, the color of solids of the primary inks and of the overprinted solids of pairs of the primary inks, and the color of the overprinted solids of three primary inks in combination with the black ink. Although color management systems can adjust data to change the outer gamut, any change in gamut requires consistency over methods for gamut compression or expansion.

If within-gamut color can be manipulated to produce matching results (as digital proofing systems do routinely) then a family of six to eight outer gamuts ranging from newsprint to high-end printing on glossy stock would define the range of printing processes that exist. Each of these gamuts would have associated with it a reference characterization data set that would be used as the transfer encoding of the color data between preparation, proofing, and printing. These would be the virtual press or reference printing conditions to which a particular gamut would be referenced. They would all be simply references between preparation, proofing, and printing. They would also be process agnostic, and represent a virtual press that was not linked to actual press performance but would be optimized for data manipulation to facilitate use of tools, such as those described in ISO 10128, to adjust the color reproduction of a printing system to match a set of characterization data.

Digital printing may fit within this family, although because it has a gamut considerably larger than ink on paper printing, it may need its own reference printing condition, or even multiple reference conditions for different types of digital printing.

References

[1] ISO (2001) 15930-1. *Graphic technology – Prepress digital data exchange – Use of PDF – Part 1: Complete exchange using CMYK data (PDF/X-1 and PDF/X-1a)*. International Organization for Standardization, Geneva.

[2] ISO (2002) 15930-3. *Graphic technology – Prepress digital data exchange – Use of PDF – Part 3: Complete exchange suitable for colour-managed workflows (PDF/X-3)*. International Organization for Standardization, Geneva.

[3] ISO (2003) 15930-6. *Graphic technology – Prepress digital data exchange using PDF – Part 6: Complete exchange of printing data suitable for colour-managed workflows using PDF 1.4*. International Organization for Standardization, Geneva.

[4] ISO (2003) 15930-4. *Graphic technology – Prepress digital data exchange using PDF – Part 4: Complete exchange of CMYK and spot colour printing data using PDF 1.4 (PDF/X-1a)*. International Organization for Standardization, Geneva.

[5] ISO (2008) 15930-8. *Graphic technology – Prepress digital data exchange using PDF – Part 8: Partial exchange of printing data using PDF 1.6 (PDF/X-5)*. International Organization for Standardization, Geneva.

[6] ISO (2001) 12647. *Graphic technology – Process control for the production of half-tone colour separations, proof and production prints*, International Organization for Standardization, Geneva.

[7] ISO (2009) 10128. *Graphic technology – Methods of adjustment of the colour reproduction of a printing system to match a set of characterization data*, International Organization for Standardization, Geneva.

18

Flexible Color Management for the Graphic Arts

18.1 Introduction

Adobe's Portable Document Format (PDF) [1,2] has become the format of choice for documents intended for print production. This chapter deals with some of the issues faced by graphic arts professionals when creating and processing PDF documents.

PDF evolved from PostScript some 20 years ago and has seen substantial development during that time to support the needs of document creators in a wide range of application areas, including document presentation on the Internet, legal documents, engineering drawings, as well as those needs of print production. Although great care has been taken by those extending the format to ensure consistency across these areas, the format has become quite complex and care must be taken when creating PDF documents if they are to be interpreted unambiguously.

A subset of PDF (PDF/X) has been defined as a series of ISO standards [3–6]. These standards enable reliable exchange of documents for print production by ensuring that all of the data that is needed to print the document is included. This format is now widely used and in part achieves its goal, but there are some areas that need further consideration, particularly those relating to the way in which the color of document elements is defined.

18.1.1 Document Preparation Objectives

One goal when preparing documents is to create a document that can be reproduced with a similar look on a range of devices. An alternative, and somewhat conflicting, goal is to define a document that can be printed accurately on a single device.

18.1.1.1 Similar Look on All Devices

This is a common requirement for advertising campaigns where material is often printed on different printing presses and is later presented together, for example, a poster and a set of leaflets. It is usually important for the brand owner that at least brand colors match in this situation.

It is increasingly common that documents must be created before the method of printing has been determined. Such documents must be created without detailed knowledge of the printing system to be used to print them.

18.1.1.2 Accurate Color on a Single Device

This is a requirement when an advert that includes a company logo or important brand color is being prepared for submission to a magazine, a company brochure is being prepared, or a home-shopping catalogue is being printed.

18.1.1.3 Achieving Both Objectives

Both of these objectives can be achieved easily if the range of colors is restricted to those colors that can be printed on all devices that will be used to reproduce the document. While this is possible and is often the solution adopted in practice, it is in many cases desirable to be able to use the full gamut of the device to be used for printing and to control the way in which different types of color elements are reproduced. PDF/X provides a framework where this goal can be realized if care is taken when constructing and rendering the document to screen and to print.

18.1.2 Describing Color in PDF Documents

There are many ways in which color can be described in PDF documents. A more complete definition is provided in the PDF specification [1,2].

18.1.2.1 Relative to Imaging Devices

Device color spaces allow document colors to be defined using the colorants of the imaging device to be used to reproduce the document.

In the case of a printing press the device colorants are usually cyan, magenta, yellow, black, and one or more spot colors. The PDF color spaces DeviceCMYK, DeviceN, and Separation allow colors to be defined relative to press colorants. The color space DeviceGray is intended for black and white devices and has a well-defined relationship to DeviceCMYK.

In the case of a monitor or similar imaging device the set of colorants is usually red, green, and blue and the DeviceRGB color space can be used to define document elements by specifying the amount of each of these colorants to be used.

Since the colorants of imaging devices vary widely, unless more is known about the device for which device colors are defined, these color spaces cannot be interpreted consistently. This means that device color spaces should never be used when creating PDF documents for print production unless further information is provided about the intended imaging device. In the case of PDF/X documents, information about the intended device is provided by the document Output Intent usually in the form of an ICC profile.

18.1.2.2 Relative to the CIE Standard Observer

Colors can be defined relative to the CIE Standard Observer using CalGray, CalRGB, and *Lab* color spaces. These color spaces are most useful where a color to be included in the document has been measured by a spectrophotometer.

Perhaps the most widely used set of color spaces is the ICCBased family of color spaces. These allow document colors to be defined relative to imaging devices but also provide sufficient information about the imaging device to allow these colors to be interpreted unambiguously.

18.1.2.3 Indirect Color

There are a number of mechanisms in PDF that allow color to be defined indirectly.

The Indexed and Pattern color spaces both provide mechanisms that allow the other PDF color spaces to be used indirectly.

Colored elements can interact with other colored elements using two mechanisms of overprinting (slightly different for process and spot colors) and Transparency.

The Rendering Intent associated with a document element modifies the behavior of ICCBased elements.

18.1.3 Document Operations

One way to think about documents is to consider the set of operations that will be performed on a document in its lifetime. For the purposes of this discussion we will consider the following subset of operations: creation, proofing, printing, and re-targeting; other document operations such as editing, merging, or re-purposing are beyond the scope of this discussion. When a document is created, ideally the set of operations that will be performed should be known as the set of data needed for each operation that must be communicated to those processing the document. A typical workflow for creation, proofing, printing, and re-targeting a document as needed is shown in Figure 18.1.

18.1.3.1 Proofing

It is usually desirable to be able to see how a document will appear when it is printed. In graphic arts print production "proofing" is often a separate step in the process. During this step a prototype print is made that accurately predicts key features of the final print as closely as possible. This includes correct simulation of all colored elements including those colored elements that are combined using overprinting or transparency effects. Increasingly "soft proofing" is used for this step and the result is simulated on a monitor whose color is carefully controlled.

18.1.3.2 Printing

At some point the document will be printed and in the simplest case the printing system for which the document was prepared will be used to print it.

Workflow Diagram

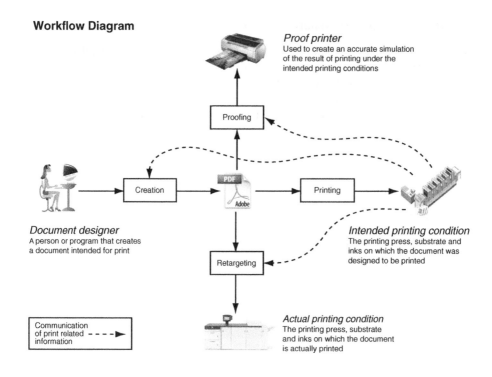

Figure 18.1   Workflow for creation, proofing, printing, and re-targeting of documents

18.1.3.3 Re-targeting

In some cases a document is prepared for printing on one printing condition but must be printed on a different printing system. This is almost always the case for today's print production as standardized reference printing conditions are used for document exchange. In many cases the actual printing system is similar to the intended printing system and uses the same set of printing inks and substrate. In these cases the printing press can be calibrated to match the standard. In other cases, particularly when digital printing presses are used, the printing characteristics are substantially different from those anticipated by the document's creator and in these cases the document must be adjusted in a more complex way to achieve a satisfactory printed result.

18.2 Requirements for Different Document Elements

We now wish to review the requirements to be able to proof, print, and re-target PDF documents. The set of data needed for these operations depends on the document content and so in this section we will look at different types of PDF content in order to understand what additional data is needed.

18.2.1 CMYK, Gray, and Black Elements

If properly qualified, DeviceCMYK and DeviceGray color spaces can be used in PDF/X documents. In some cases it is important to be able to define color relative to the printing inks in order to ensure a high-quality printed result and these color spaces can be used to achieve this.

When text is printed it is usually desirable to use black ink only, because if all four colors are used some ugly fringing appears around the text caused by slight misalignment of the printed separations.

Some pure colors look unsightly if small dark dots of another color are added and in such cases it is usually more important to keep the color pure than to keep it accurate.

Trapping is often used in order to hide the effects of misalignment of printing separations and this is often achieved by creating small elements using one or two device colorants. In many cases today the trapping stage is closely integrated with the printing system and the need to be able to communicate trapping information in this way is likely to decrease in future.

18.2.1.1 Print and Proof

Some printing systems require the total area coverage (total percentage of all ink in a region) to be limited. When creating documents that directly select CMYK amounts, care must be taken to ensure that the total area coverage limit is not exceeded, otherwise the document may not be printed correctly. This total area coverage limit must be communicated separately or guessed from the ICC profile that defines the printing condition.

18.2.1.2 Re-target

There is currently no mechanism in PDF to indicate whether pure color or accurate color is the objective for a particular element. This means that re-targeting can only be done successfully if additional constraints are imposed on the document creator and when processing the document. It is common to assume, for example, that black-only text should remain black-only when the document is re-targeted and that elements that involve only pure color should be re-targeted using the same set of colorants (although the amount of each may be changed).

If object-based re-targeting is performed to create a modified PDF document, elements that make use of overprinting or transparency may be reproduced incorrectly and so this form of document re-targeting should be avoided where possible.

18.2.2 "RGB" Images and Other "RGB" Content

DeviceRGB is not allowed in PDF/X for a document intended for print as it is not well defined. ICCBased RGB color spaces do provide a useful mechanism to allow images that were captured as RGB to be retained as RGB until the time of printing. This means that they can be converted for print using the actual press profile to be used.

In traditional graphic arts workflow images are converted for press before they are included in documents using Adobe's Photoshop or a similar application. The image's source ICC profile and the press ICC profile are used to perform an RGB-to-CMYK conversion. When performing this conversion, users select the rendering intent to be used and can choose whether or not to perform black point compensation.

PDF/X incorporates the concept of "virtual CMYK" in that image data and the transforms needed to convert them are included in the document: the source ICC profile and the press ICC profile (in the PDF Output Intent) and the rendering intent to be used. One serious limitation in PDF was that the use of black point compensation could not previously be specified in PDF – this is now supported in ISO 32000 [2].

18.2.2.1 Print and Proof

It is a serious limitation that as yet PDF/X has no way to indicate whether black point compensation should be applied. If "RGB" images are to be incorporated a separate communication between the creator and the recipient of such a document may be needed in order for a satisfactory print to be produced. A document creator could indicate, for example, that black point compensation should be applied when converting all image content. In the absence of such a communication it is only safe to assume that no black point compensation should be done.

18.2.2.2 Re-target

Re-targeting may be more difficult to achieve when ICC v4 profiles are not used as there are in some cases problems when mixing profiles from different suppliers.

18.2.3 Spot Colors

Special printing inks are often used to print company logos or to achieve a particular design effect. Such colors are referred to as spot colors and are often selected from a swatch book and identified by name. These kinds of elements are device specific as they anticipate a particular ink being available and a particular tone response of the printing system.

PDF supports this kind of color using Separation and DeviceN color spaces. These color spaces identify the color by name and provide a mechanism to convert to an alternate color space when the specified ink is not available. This alternate color definition can be used to describe how the spot color looks (using CIELAB) or to provide an alternative color (using CMYK) to be used when the spot color is not available. In some cases these two colors may be the same, but since spot colors are generally used to print colors that are outside of the color gamut of the press. they are usually different.

Since PDF only allows the definition of a single alternate color space definition, document creators must decide whether it would be more useful to define the alternate color in terms of CIELAB or in terms of process equivalent CMYK.

18.2.3.1 Print

The document creator and all those involved in processing the document must agree on the set of names to be used to define spot colors or to communicate by private means details of the ink to be used. This is often done by means of a printed color swatch book that can be referred to by both the document creator and printer. Swatch books provide a reasonably good way to communicate colors by name but have some serious limitations as they are subject to print-to-print variations and more significantly a single swatch book changes color as it ages.

It is often important to be able to specify the color of a tint of a spot color as well as the color of the solid. This information can be communicated in PDF if a CIE-based color space is used as the alternate color space. Including CIELAB color values for the solid and for each tint of a spot color provides sufficient information to allow the press operator to calibrate the press to match the colors specified by the document creator.

18.2.3.2 Proof

Proof printers usually have a significantly larger color gamut than that of the printing processes they simulate. In many cases the colors of the spot inks can also be printed by the proof printer and this means that a proof can be made that shows the result of printing including the spot colors. In order to be able to proof spot inks accurately a CIE-based color space must be used as the alternate color space for spot colors.

18.2.3.3 Re-target

In the simplest re-targeting scenario, a printing press with different tone-value increase is to be used to print the document but the document will be printed with the same set of spot inks. In this case a correction must be applied to the tint value of the spot color in order to produce the expected result on the actual press.

In many cases the spot ink is not available on the actual printing press to be used and the re-targeting system may need to perform spot-to-process conversion.

There are a number of possible objectives for this conversion. It may be important to reproduce the color as closely as possible to the original even if this means that several spot colors will end up being printed using the same process color. An alternative approach is to provide a mapping that will ensure that spot colors remain distinct when converted to process colors even if this means that these colors will show substantial variation from one printing system to another. This is very similar to the trade-off when performing other ICC color conversions, where the problem is solved using rendering intent but there is no rendering intent defined for spot color reproduction. PDF does not currently allow this additional information about the intended result to be communicated.

18.2.4 Spot and Process Color Combinations

It is quite common to combine spot colors with process colors or with other spot colors.

18.2.4.1 Print

In order to produce the correct color when printing, the printing sequence must be defined because, for example, the color produced by printing orange on top of pink is not usually the same as printing pink on top of orange. There is no way to communicate this information in PDF and so it must be communicated by private means.

18.2.4.2 Proof and Re-target

It is important to be able to estimate the color produced by the combination in order to produce a proof or to print the document on a different printing system. The opacity of each ink and the printing sequence are needed in order to be able to provide an estimate of the color combination. A popular way to measure the opacity of an ink is to print and measure patches printed on black (e.g., process black ink) and on white (e.g., the print substrate) and to use the difference in these values as a measure of opacity. Today this additional information must be communicated by private means.

It is usually impractical to create ICC profiles for each possible printing combination of spot and process colors. This means that the color produced by combinations of these colors must be defined and checked by some proprietary means.

18.2.4.3 Duo-Tones, Tri-Tones...

Multi-tones (duo-tones, tri-tones...) are defined using a small number of inks (often spot color inks) and used to print a gray image to achieve a special effect such as sepia-tone. These types of colored elements have essentially the same set of color problems as multiple spot colors; however, the possible color combinations are usually limited to a few hundred colors.

18.2.5 Transparency

Determining the color of elements that use transparency is a multi-dimensional problem.

18.2.5.1 Print and Proof

Some RIPs (not PDF/X-4 compliant) do not handle transparency correctly.

18.2.5.2 Re-target

In some cases the result of transparency blending may be slightly different due to changes in blending space.

18.2.6 Varnish

In many cases a varnish is printed as a spot ink, which has the effect of increasing the apparent color gamut so that varnished areas have high visual impact. This changes the appearance of the regions of prints that include the varnish. There is no way in PDF to communicate the additional information needed to be able to produce an accurate simulation of the final printed result and so, if it is important to be able to simulate the effect of varnish, additional information must be communicated.

18.2.6.1 Print

Varnish inks must be clearly identified. There is no standard way to do this and so varnish inks are usually just identified using a name that includes the word "varnish."

18.2.6.2 Proof

The way in which varnish works is to change the surface finish of a print, usually making it more glossy. Unless a varnish ink is available for the proofing system it is not really possible to simulate the effects of varnish on a printed proof. One of the benefits of soft proofing is that the effects of gloss can be shown, but since there is no standard way to communicate the data needed to provide an accurate simulation of the effects of adding the varnish, this additional data must be communicated by private means.

18.2.6.3 Re-target

Unless a varnish ink is available on the actual printing system to be used some redesign of the document is usually necessary so that the needed impact can be achieved by some other means.

18.2.7 Special Inks

For some printing applications, especially in the area of packaging printing, special effects inks are used, such as metallic, fluorescent, and pearlescent inks. Accurate modeling of the appearance of such inks is an area of active research; however, there are a number of models that predict the results to a useful degree.

18.2.7.1 Proof

Models that allow the appearance of these inks to be predicted require knowledge of the lighting and geometry of the viewing environment as well as multi-dimensional measurements of the ink characteristics. This means that with the exception of a few critical cases, it is impractical to derive and communicate the data that is needed to allow accurate simulation.

Some monitor proofing applications can simulate effects of special inks but no standard means exists to communicate the additional metadata.

18.2.7.2 Re-target

There is really no way to provide a reasonable approximation to the appearance of these special inks using process colors and so, unless the intended ink is available, some redesign of the document is needed.

18.2.8 Putting the Elements Together

If care is taken when creating them, most kinds of PDF documents can be proofed, printed, and re-targeted reliably, but in most cases additional information about some of the document content must be communicated by separate means.

The separate means can be basic industry rules, for example, black-only text elements should remain black-only when the document is re-targeted, or maybe additional measurement data is needed to define the color and opacity characteristics of spot colors. It is important that both the creator of the document and those responsible for subsequent processing know and understand the implications of the basic assumptions made. The limitations of PDF for data exchange, and particularly the problem of representing black-only and spot color elements, are currently being addressed in working groups in ISO TC 130 and ISO TC 171.

References

[1] Adobe Systems (2008) Document management – Portable Document Format – Part 1: PDF 1.7.

[2] ISO (2008) 32000-1. Document management – Portable Document Format – Part 1: PDF 1.7.

[3] ISO (2003) 15930-4. *Graphic technology – Prepress digital data exchange using PDF – Part 4: Complete exchange of CMYK and spot colour printing data using PDF 1.4 (PDF/X-1a).*

[4] ISO (2003) 15930-6. *Graphic technology – Prepress digital data exchange using PDF – Part 6: Complete exchange of printing data suitable for colour-managed workflows using PDF 1.4 (PDF/X-3).*

[5] ISO (2008) 15930-7. *Graphic technology – Prepress digital data exchange using PDF – Part 7: Complete exchange of printing data (PDF/X-4) and partial exchange of printing data with external profile reference (PDF/X-4p) using PDF 1.6.*

[6] ISO (2008) 15930-8. *Graphic technology – Prepress digital data exchange using PDF – Part 8: Partial exchange of printing data using PDF 1.6 (PDF/X-5).*

Part Four

Measurement and Viewing Conditions

19

Standards for Color Measurement and Viewing

Until recently, the standards for viewing conditions, colorimetric measurements, densitometric measurements, and characterization data for profile building were all independent of each other and in some cases there were inconsistencies between them. The standards that define viewing conditions and densitometric measurements in particular were historic standards that had not caught up with the needs of color management and the realities of the current market.

This situation has now changed with the revision of the key standards for measurement and viewing. Although the viewing and densitometric standards are primarily the responsibility of ISO TC42 (Photography), and the colorimetric measurement standards are under ISO TC130 (Graphic technology), several joint working groups have been set up to revise these standards in concert with each other. At the time of writing, these standards are in final ballot or being prepared for publication by the ISO, and when adopted the result will be far better consistency between the various requirements.

19.1 The Driving Forces for Change

Although these standards are often used completely independently of each other, the issues driving the need for coordination largely come from the requirements of the color management and graphic arts production areas which use all of them in a coordinated manner. The following are some examples of the diverse issues facing the community of technical experts that tackled the revision of these documents:

1. With the move to widely available spectral reflectance measurement equipment the industry would like to be able to compute both density and colorimetry from the same spectral data. This meant that a spectral definition of density was needed. In addition, provision needed to be made for a common illuminant spectral power distribution (SPD) in the measurement equipment and a common backing to be used under the specimen being measured.

Color Management: Understanding and Using ICC Profiles   Edited by Phil Green
© 2010 John Wiley & Sons, Ltd

2. For color management, it was desirable to be able to correlate what was seen in the viewing booth with the colorimetric measurements made of the proof and print being compared. Here the issue was to be sure that the SPD of the viewing booth was an adequate match to the SPD used in the illumination system of the spectral measurement equipment and that the illumination used for the computation of colorimetry also matched. In many situations the backing used for both measurement and viewing was also an issue.

3. The amount of optical brightening agents (OBAs) used by the paper industry to make paper appear brighter and whiter has increased in recent years in response to customer demand for brighter and "whiter" paper. These agents absorb energy in the ultraviolet and emit in the blue portion of the visible spectrum. This makes the SPD of the illuminants (and in particular the amount of UV they contain) used in measurement equipment and viewing booths more critical.

4. Although for years the European graphic arts community has used polarization filters for density measurements, and to a lesser extent colorimetric measurements, the standards have not made any provision for defining polarization.

5. Although the SPD of a standard illuminant can be specified, the tolerances on the compliance of a real illumination source used to simulate such an illuminant must be specified in terms of their effect. Thus, tests like chromaticity, color rendering index, metameric index, etc., laid out in various CIE and ISO documents, must be used for evaluation. The computational requirements of these tests are somewhat difficult to follow and ensure precise compliance, so it was considered useful to include an annex to the viewing standard that laid out a step-by-step example calculation of all of the various parts of the specified evaluation procedure.

19.2 The Standards

The standards involved in graphic arts measurement and viewing are:

ISO 5, *Photography – Density measurements*
  Part 1: Terms, symbols and notations
  Part 2: Geometric conditions for transmittance density
  Part 3: Spectral conditions
  Part 4: Geometric conditions for reflection density
ISO 3664, *Viewing conditions – Graphic technology and photography*
ISO 13655, *Graphic technology – Spectral measurement and colorimetric computation for graphic arts images*

The ICC specification, ISO 15076-1, *Image technology color management – Architecture, profile format and data structure – Part 1*, references these specifications and refines the viewing conditions and profile measurement conditions for color management.

19.3 Making the Standards Work Together

There are many places where one could start in describing the interaction and overlap resulting from this latest set of revisions, but in this chapter the intention is to simply identify key areas and show how they are handled.

19.3.1 Backing

The driving force behind the backing issue is that traditionally densitometry has used a black backing to minimize the effects of printing or other marks on the back side of the specimen being measured. However, in color management, when prints and proofs are being measured colorimetrically, black backing will darken the appearance of a thin or translucent sheet. If all proofs were made on the same substrate as was used for printing, that could minimize the problem. However, most proofing materials are more opaque than the printed sheet that they are simulating and thus less sensitive to the backing showing through. A white backing allows the substrate of the printed sheet and that of the proof to more closely match.

Thus if color management profiles are to be used to help make proofs that match printing conditions, colorimetric measurements must be made with a white backing.

The apparent conflict between black and white backing requirements was resolved by allowing both in ISO 5-4, ISO 13655, and ISO 3664, with the requirement that the backing used be identified. In addition, in ISO 13655 a procedure is given to modify tristimulus data for half-tone images for different substrate reflectances such as those caused by differences in backing materials. This correction method is also described later in this document.

19.3.2 Computation of Density from Digital Data

Prior to this revision of the ISO 5 series of standards, ISO 5-3 made no provision for the computation of density from digital data. It defined the spectral response functions that a filter densitometer was required to meet when the illumination system, the optical system, the detector, and any necessary filters were combined. The required response was defined at 10 nm intervals.

This revision of ISO 5-3 interpolated this data to a 1 nm interval. At 1 nm the spectral response function and spectral weighting factors are the same. However, once the data interval is increased to 5 or 10 nm, or even greater intervals, the spectral weighting factors must include both the densitometric spectral products and the coefficients of a polynomial for interpolating the spectral reflectance factor or transmittance.

These spectral weighting factors allow the computation of density directly from spectral reflectance (or transmittance) data. The standard also describes how to compute the spectral weighting factors for any data interval. These computations assume that the illumination system for measurement is CIE illuminant A.

19.3.3 Polarization

Prior versions of ISO 5-4 did not include any provision for the use of polarization in making density measurements, in spite of the fact that polarization is widely used in Europe and other parts of the world to minimize differences between wet and dry measurements. This revision of ISO 5-4 adds polarization and defines a requirement and test for polarization efficiency.

While polarization is not common in colorimetric measurements and not very practical in viewing, a designation for that type of illumination is provided in all three of the standards. This is discussed further in the section on illumination sources that follows.

19.3.4 Illumination Sources

Perhaps the most significant change in ISO 13655 relates to the issue of illumination conditions and their applicability. These conditions are now fully defined in ISO 13655 and referenced from the other standards. While not all of the illumination conditions are applicable in all of the standards, provision is made in each standard for identifying the condition used. The key requirement that needed to be addressed is the need for consistency in measuring optically brightened paper. The other issue was compatibility between metrology and viewing, which again is a significant issue for optically brightened papers. The four conditions specified are as follows:

1. Measurement condition M0. Spectrophotometers used in the graphic arts have incorporated an incandescent light source (with a relative SPD that is close to CIE standard illuminant A) or more recently LED sources. M0 is an undefined SPD, and is intended to be used to give provenance to existing spectral data.
2. Measurement condition M1. To minimize the variation in measurement results between instruments due to fluorescence (by optical brighteners in the substrate and/or fluorescence of the printing and/or proofing colorants), the SPD of the light flux incident on the specimen surface for the measurement should match CIE illuminant D50. There are two methods provided to achieve conformance to condition M1:
   (a) The SPD of the measurement source at the sample plane should match CIE illuminant D50. Conformance is based on the criteria specified in ISO 3664, including the metamerism index requirements for the UV region.
   (b) A compensation method is used in combination with a controlled adjustment of the radiant power in the UV spectral region below 400 nm. This can be done by active adjustment of the relative power in this range with respect to a calibrated standard for D50. This compensation aims only to correct the effects of fluorescence of optical brighteners in the substrate. The SPD in the range from 400 to 700 nm should be continuous.
3. Measurement condition M2. To exclude variations in measurement results between instruments due to the fluorescence of optical brighteners in the substrate surface, the source should only contain substantial radiation power in the wavelength range above 400 nm, with a longpass or "UV cut" filter used to suppress radiation below 400 nm. The spectral power of the instrument source is otherwise unspecified.
4. Measurement condition M3. For use in special cases, an instrument may be equipped with means for polarization in order to suppress the influence of first-surface reflection on the color coordinates. An instrument fitted with a polarization filter must also fulfill the requirements of measurement condition M2.

It is important to realize that when measuring specimens that have no OBAs and do not use fluorescent inks, measurement conditions M0, M1, and M2 should all produce identical results.

19.3.5 Validation of D50 SPD

ISO 3664 specifies four criteria (and associated tolerances) for evaluating the adequacy of the conformance of an illumination source to CIE D50. These are:

- Chromaticity aim and tolerance
- Color rendering index (general and special indices)
- Metamerism index (visible and UV indices)
- UV metamerism index.

Although these are all defined in referenced CIE publications, from a practical point of view the varying subtleties between the documents and test can be very confusing. This can lead to small disagreements between the calculations made by different laboratories. To try to help this situation, the informative Annex D has been added to ISO 3664. This annex provides a detailed step-by-step review of the required computations and pointers to the specific equations to be used. It also provides worked examples of each by utilizing CIE illuminants D55 and F8 as test sources to simulate the reference D50 illuminant. These were chosen, not because they are valid substitutes for D50, but because their SPDs are documented in appropriate CIE publications and therefore are readily available and will be consistent for any user. The intent is that anyone developing evaluation software can use this data to test and validate their computations and increase the probability that computations from different laboratories will match.

19.3.6 Color Characterization Data

ISO 15076-1 points to ISO 13655 as the basis for the colorimetric data in characterization data sets, but notes that white backing is preferred for color management applications. As ISO 13655 requires that the backing used be specified, this enables the use of the tristimulus correction method, where necessary, to convert data measured on other than a standard white backing to be converted to a white backing equivalent.

19.3.7 Tristimulus Correction Method

Although the tristimulus correction method preceded the work involved in bringing these standards into agreement, it did play a significant part in enabling the vision that this coordination was possible. It is therefore included here for general information.

The tristimulus correction method is based on the observation that when the differences in X, Y, and Z between measurements made over white and black backing materials are plotted against the X, Y, and Z values for measurements made over either material, the best-fit result is approximately a straight line. At the lowest value of each tristimulus value, the difference between measurements made over the two backings is at or near zero. The maximum difference in measurement due to backing material characteristics is always at the maximum tristimulus value which equates to a measurement of the unprinted substrate alone.

This implies that measurements over one backing can be used to estimate the measurements that would be made over another backing by simply adding a correction factor in X, Y, and Z. This correction factor is simply a proportional amount of the difference between measurements of the substrate alone over the two backings where the proportion added is defined by the value of X (or Y, or Z) on the first substrate compared to the minimum value of X and the value of X for the substrate alone. This leads to a correction equation for X as follows:

$$X(n)_2 = X(n)_1 + (X(s)_2 - X(s)_1) \frac{(X(n)_1 - X_{\text{MIN}})}{(X(s)_1 - X_{\text{MIN}})}$$

where:

$X(n)_1$ = Measured value of X for sample n over backing 1;
$X(n)_2$ = Predicted value of X for sample n over backing 2;
$X(s)_1$ = Measured value of X of the substrate over backing 1;
$X(s)_2$ = Measured value of X of the substrate over backing 2;
X_{MIN} = Minimum value of X which generally corresponds to a four-color solid, which is patch ID 24.

The corrections for Y and Z are performed in the same way, These new X, Y, and Z values are then used to compute new CIELAB values.

19.4 Looking to the Future

These standards are living documents within the ISO TC130 and TC42 committees and will certainly be revised again in the future. It is the responsibility of the TC130, TC42, and ICC and the industry groups they represent to monitor these documents to both initiate needed changes and ensure that any future changes correspond to the best practice at the time and are reflected across all documents.

20

ICC Recommendations
for Color Measurement

20.1 Introduction

In order to achieve the highest quality results in color management systems including (and not limited to) those using profiles constructed according to the ICC standard, it is essential to measure color accurately and consistently.

CIE, the International Commission on Illumination, has published various recommendations for color measurement, as well as calculation procedures, that can be used for this task. The calculation procedures have been extended in various ISO and ASTM standards. So, users of colorimetry are often confronted with various measurement recommendations which may seem appropriate for their task, each of which could give rise to a different colorimetric result.

It is important, then, that measurement procedures be more uniquely and unambiguously identified, to ensure consistency between users and instruments. This White Paper summarizes the issues that users should consider when making color measurements for the purpose of constructing ICC profiles, and describes the recommended practices.

20.2 Reflecting Media

The ISO 13655 standard specifies how color measurements and calculations for use in graphic technology are to be conducted. This specification is the basis of the measurement procedures specified by the ICC. ISO 13655 requires that instrument geometry be either 0:45 or 45:0 and that all calculations of tristimulus values be achieved using the CIE 1931 Standard Colorimetric Observer. ISO 13655 also calls for using CIE illuminant D50 for the computation of colorimetric values and defines spectral weighting functions derived from this observer and illuminant. These weighting functions should be used when measurements are made with a spectrophotometer or spectroradiometer in which the spectral sampling interval is coarser than that specified by CIE, that is, less than or equal to 5 nm.

Color Management: Understanding and Using ICC Profiles   Edited by Phil Green
© 2010 John Wiley & Sons, Ltd

The previous version of ISO 13655, published in 1995, recommended that the source of illumination conform to D50, at least when fluorescence is present in the sample, and that a black sample backing be used when making measurements. These requirements presented some difficulties when making measurements intended for use in developing the characterization data required for the construction of ICC profiles. For ICC color management to work correctly, colorimetry should be measured using the illumination and backing that will be used in practice, when the reproductions are viewed by an observer.

ISO 13655:2010 defines multiple measurement conditions for different applications. These conditions are referred to as M0, M1, M2, and M3, and are described further in Chapter 9. It also permits the use of a white sample backing. This White Paper seeks to clarify how the measurement conditions defined in ISO 13655:2010 should be applied for profile making purposes.

20.2.1 Fluorescence

Many of the instruments likely to be used in the graphic arts workplace rely on tungsten or tungsten–halogen light sources. Although these lamps emit substantially less ultraviolet radiation than is specified by the D50 illuminant, they will still cause excitation of fluorescing materials.

Instrument manufacturers often offer the opportunity to remove this incident UV radiation. The experience of ICC members suggests that it may be beneficial to take this opportunity when making measurements for profiling, even though removing the UV may cause differences from the color perceived in a typical viewing environment.

The ICC bases this recommendation on two factors. First, the UV excitation in measurement may be different from one instrument to another, as well as varying from that in the source of illumination used for viewing the print. For this reason the measured stimulus can only be an approximation that may, or may not, produce a color measurement closer to the perceived print color. Given this uncertainty, greater consistency between measurements may be achieved if the UV is excluded.

The second factor arises from the fact that for most printed media the strongest fluorescence is found in the substrate. In a proofing situation, the proof medium and the print medium may have very different fluorescence. If relative colorimetry is used, which normalizes the data to the substrate, the normalization procedure can introduce unacceptable differences between the ink colors. The normalization is avoided by using the ICC-absolute colorimetric rendering intent, but this choice can lead to unacceptable highlight clipping. Often the best choice is to reduce the measured difference in the substrates by excluding the UV.

In practice, most indoor viewing environments in which color reproductions are viewed outside of the viewing booth have relatively little or no UV present in the illumination.

It should be noted that UV filters absorb light below a certain wavelength, typically around 400 nm. However, the filters used for this purpose do not have a perfectly square cut-off and hence permit a proportion of UV below 400 nm to reach the sample. Moreover, fluorescing materials are excited to some degree by light energy above 400 nm. Consequently, exclusion of the UV may not completely eliminate fluorescence effects.

Also, UV-excluded measurements of fluorescent materials will not correspond well with visual observations when using a viewing booth which simulates the D50 spectral power distribution in the UV region. In this situation, the ideal is to have the spectral power distribution of both the viewing booth source and that of the source used for the measurement correspond to CIE illuminant D50.

In this situation the ISO 13655:2010 measurement condition M1 is appropriate. This measurement condition allows media-specific compensation techniques to be applied by the instrument manufacturer so that the actual source used in the instrument does not have to match the spectral power of the D50 illuminant, although some UV must be present.

Since measurements made according to M1 and M2 conditions will usually differ when measuring fluorescent samples, characterization data should always specify which condition was used. Profiles made from M1-measured data should ideally be evaluated through measurements made using an M1-conforming instrument, and similarly profiles made with M2-measured data should be evaluated through measurements made with a UV-excluding instrument.

M1 measurements may not be appropriate for profiling purposes where there are significant differences in fluorescence between the substrates used for proofing and printing.

Chapter 21 has more information on this topic.

20.2.2 Backing

ISO 13655:2010 allows both white and black backings when making color measurements for graphic arts. The effect of a black backing on a substrate that is not completely opaque is to reduce the color gamut (including the dynamic range). Although it can be argued that a black backing may in some circumstances be a better match to the way that a print is intended to be viewed, the experience of many ICC members is that it is better to use a white backing for measurements. Hence, ICC recommends the use of white backing when measuring samples for making profiles.

The full description of the recommended white backing material is given in ISO 13655:2010, but essentially it should be opaque, diffusely reflecting, and with a C^* chroma value not exceeding 2.4.

20.2.3 Polarization Filters

Some measuring instruments allow the user to polarize the incident beam to reduce the effect on the measurement of gloss changes following printing, as specified in ISO 13655:2010 measurement condition M3. In such instruments, a second polarizer is placed in the reflected beam in an orthogonal or "crossed" orientation and can largely remove the specular reflections that change their characteristics as inks dry. While including this specular component may provide a better indication of the perceived final result, it can present problems in process control where measurements may be compared between prints where the ink is dry and those where it is still wet to some degree. The use of polarization filters largely eliminates the measurement differences that can arise between such samples, which can be beneficial in process control. However, since the properties of such polarization filters are undefined and vary between instrument models, the use of such filters tends to result in a reduction in inter-instrument agreement, in addition to decreasing the correspondence with visually observed results.

The ICC recommends that when making measurements for the purpose of generating profiles, polarization filters should be removed, and instruments with polarization that cannot be removed should not be used. (Instruments with polarization may still be appropriate for in-plant process control purposes, to monitor the deviation between dry proof and wet print.)

Furthermore, time should where possible be allowed to ensure that the print is properly dry before making measurements. In some cases, the drying time required to stabilize color measurements is longer than might be expected from just observing the surface gloss or tackiness.

20.3 Measurement and Calculation Procedures for Transmitting Media

The recommendations of ISO 13655 should be followed when measuring transmitting media, with the exception of the source of illumination, which is generally less critical because transmitting substrates with fluorescence are extremely uncommon.

ISO 13655 specifies that the measurement geometry for transmitting media should be either 0:diffuse or diffuse:0. If an opal glass diffuser is used, it should conform to that defined in ISO 5-2. The procedure for the calculation of tristimulus values should be the same as for reflecting media, by using the CIE 1931 Standard Colorimetric Observer (2°), with the CIE illuminant D50. The ISO 13655 spectral weighting functions, derived from this observer and illuminant, should be used when the measurement is made with a spectrophotometer or spectroradiometer in which the spectral sampling interval is coarser than that specified by CIE – that is, less than or equal to 5 nm.

Of the different ISO 13655 measurement conditions, ICC recommends an M2 condition (typically achieved with a tungsten source conforming to that in ISO 5-2), with any UV excluded, when making measurements for characterization data intended for the creation of ICC profiles.

The recommendations as to averaging a number of measurements should be consistent with those recommended for reflection media, except where the image being measured is a commercial input target, in which case the issues of consistency and uniformity should be unimportant as the target should not exhibit such problems.

20.4 Measurement and Calculation Procedures for Color Displays

ISO 13655:2009 addresses the measurement of self-luminous sources, such as color displays. Many other standards or recommendations also do so, including CIE Publication 122, IEC 61966 (parts 3–5), and the ASTM standards E1336 and E1455. These specifications recommend measurement procedures as well as measurement instrument characteristics. Among them they cover measurements obtained with both spectroradiometers and tristimulus colorimeters. Measurements of displays should be consistent with the recommendations made in the standards appropriate to the type of display and/or measurement device used. If the measurement instrument is in conformance with these standards, then the user need address only a relatively small number of issues.

Care should be taken when making measurements to ensure that the sampling frequency, or integration time, of the instrument used is synchronized with the frequency of scanning of the display. If not, at least 10 measurements should be taken and averaged.

Although the use of telespectroradiometers or telecolorimeters for measurement from the viewer position is often advantageous, they are not in common use among those building profiles. The ICC recommends that they be used whenever possible for display measurements, as they will include any veiling glare present, and therefore provide an accurate representation of the color as perceived by the viewer. Where such instruments are not available, and measurements are made in contact with the face of the display, some attempt should be made to measure the veiling glare from the viewer position, so the result can be used to correct the contact measurement data obtained. If a telespectroradiometer or telecolorimeter is not available, a spot light meter can be used to get the approximate ratio of the luminance of the display faceplate, as observed from the viewer position, with the ambient illumination on and off. This ratio can be used to estimate the veiling glare from the display black contact measurement. The contact measurements are corrected by adding the veiling glare to them, typically assuming that the veiling glare has the same chromaticity as the display white point for simplicity. If it is not possible to obtain any estimate of the veiling glare, the contact measurements should be corrected by assuming a veiling glare of $1 \, \text{cd/m}^2$. However, users should be aware that this level of glare may not be correct for their specific viewing conditions, which is why the two previously described methods are preferred.

Where display profiling software allows users to specify the veiling glare as part of the input for profile construction, the software should perform the data correction. When this is not the case, the user will have to correct the data prior to building the profile.

It should be noted that, in this context, veiling glare refers to the ambient light reflected from the display faceplate in the direction of the viewer. It does not refer to flare internal to the display, which should be included in contact measurements if measurement patches are displayed with an appropriate surround. It also does not refer to any flare that may result from ambient illumination not from the display entering the measuring instrument or eye, as this type of flare is not supposed to be included in profiles and, if present, should be removed from measurement data before it is used for profile construction.

Measurements of the display should be made to ensure acceptable levels of constant channel chromaticity, spatial uniformity, internal flare, and channel independence. Those displays exhibiting poor uniformity or high levels of internal flare should be avoided, or care taken to average measurements made with varying image surround and/or position. For displays with inconsistent channel chromaticities, or low channel independence, profiles should be based on an n-component LUT rather than a three-component matrix.

When spectral data is obtained during measurement, the CIE 1931 Standard Colorimetric Observer ($2°$) should be used for the calculation of tristimulus values. Spectral data should be obtained at wavelength sampling intervals of no more than 5 nm. In some cases finer sampling intervals will be required to obtain sufficient colorimetric accuracy, as some display primaries exhibit narrow spikes in their spectral radiance which are not well captured in an instrument with a wider interval.

When using a telespectroradiometer, measurements should be taken from a display area of at least 4 mm in diameter with an angle of collection of $5°$ or less. Averaging to avoid measurement errors should also be undertaken.

20.5 Number of Measurements

Two significant issues must be addressed when making measurements for the construction of profiles:

- Device consistency and uniformity
- Errors during measurement.

Averaging multiple measurements can minimize the impact of both factors.

A profile is appropriate for the condition obtained by the calibration of the device at the time when the profiling target was printed. But for many devices, however carefully they are calibrated, some variation will occur over time. The ideal profile should as far as possible reflect the central value within this variation, minimizing its effect by averaging multiple measurements.

Some printers, particularly offset printing presses, can suffer from a lack of uniformity over the sheet. In part, this is caused by the ink coverage in other parts of the sheet. In an attempt to minimize the effect of this variation, some profiling targets are "randomized" to avoid relatively large areas of each ink being localized on the print. The ICC recommends the use of randomized targets, if available. When they are not available or when the potential printed area is much larger than the target, measurements should be made of multiple targets taken from different positions on the sheet, with various orientations of the target. These should be averaged to obtain the data to be used for profiling.

Errors may arise during measurement, due to measurement technique or poor instrument repeatability. To minimize the effect of these errors, the ICC recommends that the average of a number of measurements of each patch of the target be used when making profiles.

These are recommendations for the "ideal" situation. How many measurements need to be averaged depends on the consistency and/or uniformity of the device, the instrument repeatability, and/or the competence of the operator. Prior knowledge of the significance of these factors may permit single measurements to suffice – however, without that knowledge multiple measurements should be averaged as described here.

An advantage of basing profiles on well-prepared measurement data, which result from averaging multiple printed samples and multiple measurements, is that the forward and inverse transforms tend to be significantly more accurate.

20.6 Summary of the Recommendations

The recommended measurement conditions and procedures described above are summarized below:

- Reflectance and transmittance measurements of non-fluorescent media should conform to ISO 13655:2009 measurement conditions M1 or M2. The exception is when the actual illumination will be significantly different from D50. In this case, the profile construction should use the colorimetry corresponding to the actual illumination. (As noted in Chapter 19, historic characterization data may be considered to be ISO 13655:2009 measurement condition M0.)

- In certain situations, where the end-use viewing condition includes a significant amount of UV and the substrates used fluoresce, the ISO 13655:2009 M1 condition, in which the measurement source effectively matches CIE illuminant D50, should be used.
- The use of M0, M1, or M2 measurement conditions should be reported when exchanging measurement data or profiles made using such data.
- For reflectance measurements a white sample backing is recommended.
- For reflectance instruments the use of polarizing optics should be avoided.
- For displays, measurements should conform to ISO 13655:2009. Additionally, display measurement instruments should be consistent with the recommendation of CIE Publication 122, IEC 61966 (parts 3–5), or the ASTM standards E1336 and E1455. Measurement should ideally be made with a telescopic instrument at the viewer position, but where this is not possible, and the measurement is made using an instrument in contact with the face of the display, the veiling glare at the viewer position should be measured. If this cannot be done, a veiling glare of 1 cd/m^2 should be assumed.
- When contact measurements are made of displays, the veiling glare should be used to correct the data prior to profile construction, unless profile building software allows this as a separate input. Multiple measurements should be made to minimize the effect of poor synchronization between the display scanning frequency and measurement integration time.
- For all media, multiple measurements of each patch should be averaged. The extent of this should be consistent with the uniformity and/or temporal consistency of the device, and temporal consistency of the measurement instrument and/or operator.

21

Fluorescence in Measurement

Most commercial printing papers on the market have significant amounts of fluorescent whitening agents, or FWAs (also known as optical brightening agents, or OBAs), to maximize their whiteness and brightness. These additives are important in producing modern, highly brightened papers in response to customer demand.

FWAs contain stillbene molecules that are excited by photons in a spectral band that lies mainly in the UV, and in response emit photons in a band which lies mainly within the visible spectrum. The excitation and emission regions peak at approximately 350 and 440 nm respectively.

Measurement of fluorescing materials is not straightforward. Colorimetric measurements of color prints are derived from measurements of the reflectance factor, which is the ratio of the reflected radiance to the radiance reflected under the same conditions by a perfect reflecting diffuser. Since this ideal diffuse reflector is non-fluorescing, the regular component of the total reflected radiance is also free of fluorescence. However, the human visual system (and most measurement systems) also responds to the fluorescent radiance component if present in the reflection, and does not distinguish between regular and fluorescent components.

While the regular radiance component of the measurement can readily be calibrated so that it is independent of the source illumination, the fluorescent radiance component is dependent on the amount of energy emitted by the instrument source within the excitation region. A range of different sources are used in graphic arts instruments, including tungsten, pulsed xenon, and LEDs, and it is difficult to obtain good inter-instrument agreement and repeatability between all types of instrument.

Many instruments suppress energy in the excitation region through the use of longpass filters commonly referred to as UV-cut filters. However, the suppression of excitation energy cannot be achieved in an ideal way by the use of such filters, since they have some transmission in the excitation band and some absorption in the visible band; moreover, the two bands overlap over the region 380–420 nm, so that complete suppression of excitation energy would lead to a loss of response in the blue end of the spectrum. A complete measurement of the fluorescent component of reflection can only be achieved by a bispectral instrument.

Color Management: Understanding and Using ICC Profiles Edited by Phil Green
© 2010 John Wiley & Sons, Ltd

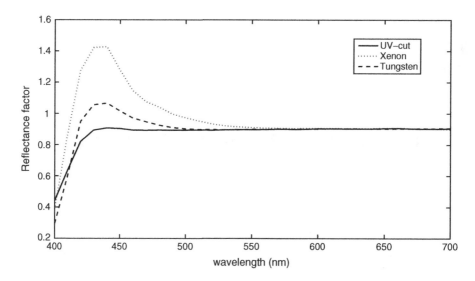

Figure 21.1 Spectral reflectance of white paper measured using xenon, tungsten, and UV-cut sources

Figure 21.1 illustrates a highly brightened white printing paper measured with xenon, tungsten, and UV-cut sources. The UV-cut source is in effect an ISO 13655 M2 measurement condition, while the tungsten source corresponds to an ISO 13655 M0 measurement condition. The xenon source has a relative spectral power in the UV excitation region similar to D50, and so is closer to the ISO 13655 M1 measurement condition, while not matching it within the tolerances defined in ISO 13655. Table 21.1 shows the CIELAB values arising from the three reflectances, together with the CIELAB ΔE_{ab}^* difference relative to the UV-cut measurement.

Measurement of FWA-containing substrates is further complicated because FWA efficacy decreases on prolonged exposure to UV radiation.

A CIE study [1] of UV-excluded and UV-included measurement of printed samples, using an instrument with a xenon source, found differences of the order of 12 ΔE_{ab}^* for unprinted paper and 3–4 ΔE_{ab}^* for solid inks, on a highly brightened paper. The largest differences are found in unprinted paper and lighter tints, while darker tints mask the fluorescence somewhat. Where present, yellow ink tends to absorb UV radiation effectively and minimize fluorescence.

A viewing booth conforming to ISO 3664 is required to match the CIE D50 illuminant in the UV as well as the visible. The D50 illuminant is defined over the range 300–800 nm, and has a significant amount of UV content, which is not matched spectrally by the D50 simulators used in commercial viewing booths. Moreover, end-use viewing environments have varying amounts of UV depending on the type of lamps used and the permittivity of window glass.

Table 21.1 CIELAB values for measurements of white paper in Figure 21.1

	L^*	a^*	b^*	ΔE_{ab}^*
UV-cut	95.96	0.14	0.71	0
Tungsten	96.09	2.10	−5.24	6.27
Xenon	96.94	5.18	−17.70	19.11

This degree of uncertainty in measurement and viewing poses a number of problems in color management. First, the measurement of the sample depends on the UV in the instrument source, but the appearance depends on the UV in the viewing illumination, and these may not be well matched. Secondly, different media often have different amounts of FWA and, where this is the case, matching the white point spectrally is difficult or impossible. In addition, any apparent visual match between media with different amounts of FWA will only hold under one viewing condition.

Color management operates on colorimetric coordinates, and, on a reflective medium, increasing the peak reflectance is not possible. As a result, the closest colorimetric match (in a minimum ΔE_{ab}^* sense) is achieved by a color with a larger negative b^* value, resulting in a more bluish rather than a whiter appearance.

Recent revisions to the ISO standards for graphic arts measurement and viewing conditions (ISO 13655:2009 and ISO 3664:2009) provide two possible approaches to the problem of matching proof to print with FWA-containing substrates:

1. Discount the fluorescent radiance by excluding UV from the measurement source, using measurement condition M2 in ISO 13655. This will eliminate most of the uncertainty which arises from fluorescence, and will also tend to lead to more similar colorimetric values for the media white on both brightened and unbrightened papers. This approach is appropriate when there is little or no UV in the end-use viewing condition, but if the proof and print media have different amounts of FWA they will not match when compared in a viewing booth conforming to ISO 3664.
2. Ensure that the amount of UV in both measurement and viewing conditions is matched, using measurement condition M1 defined in ISO 13655, and viewing prints in a booth whose light source simulates D50 in the UV as well as the visible, within the tolerances defined in ISO 3664. This approach is applicable when there is a significant amount of UV in the end-use viewing condition.

The ICC recommends the first of these two approaches in most situations, except where there is a significant amount of UV in the end-use viewing condition and the measurement instrument has an M1 measurement condition.

Chapter 20 provides more information on the measurement of imaging media for color management.

References

[1] CIE (2004) *The Effects of Fluorescence in the Characterization of Imaging Media*, Publication 163:2004, Central Bureau of the CIE, Vienna.

22

Measurement Issues and Color Stability in Inkjet Printing

It has been observed that inkjet prints exhibit color change following printing. This can be an issue in situations where color accuracy is critical, such as proofing. Profiles produced from measurements of inkjet-printed test charts may not describe a stable state of ink and media interaction, and prints which are within a given tolerance when printed might change to the extent that they are no longer in tolerance when appraised.

The aim of this chapter is to describe the common types of inkjet paper media and their performance with dye-based and pigment-based inks presently on the market, and to indicate the magnitude of color shifts which can be experienced.

22.1 Inkjet Media

The basic media types are: uncoated, matt coated, gloss coated, swellable, and microporous. These categories do have several variations thanks to the manufacturers' efforts to improve product performance and reduce costs.

The uncoated media type is the basic surface-sized paper. While the manufacture will often be to a high standard, the performance is inferior to coated media in terms of color and image quality and therefore will not be considered any further here.

The aim of the paper coating is to give the optimum color strength and dot definition to give the optimum image quality with the quickest drying time. Therefore the dye or pigment has to stay at or close to the surface while the ink vehicle has to be drawn down and dispersed into the bulk of the coating and paper. How this is done depends on the coating type. What has been found is that the color formed is not stable even under standard room conditions.

For matt coated papers, the ink is jetted onto a filled coating containing a high proportion of silica mixed with other fillers and pigments (e.g., calcium carbonate and titanium dioxide) bound with polyvinyl alcohol (PVOH). The dye or pigment will be electrostatically attracted to the silica, and so will remain at the surface. But each dye will be attracted to the silica to a varying degree according to the type of dye molecules present. This can lead to migration further into the layers, especially when surface silica particles receive a large volume of ink. To help stabilize this situation, dye fixants or mordants can be mixed with the PVOH to restrict the dye movement, though this can cause problems with removing the ink vehicle. The ink vehicle needs to travel past and disperse through the PVOH/silica coating. Dye fixants (and any other performance additives present) may impede the removal of the vehicle and actually allow the dye to move around. Another factor that can occur is a change in color due to dye and dye fixant interaction. This may change over time with the changing ratio of free and bound dye molecules, and is more of a problem with cheaper papers. Similar electrostatic interactions occur with pigment inks, but the aim is to allow proper orientation of the pigment on the surface so the control of how quickly the ink vehicle is removed is vital and the surface color can change while the pigment dries.

Glossy coated papers can be similar in structure to the matt coated papers but tend to have at least two coated layers over a very smooth paper coated with clay or barium sulfate. In the top layer a lower volume of fillers is used, with functional polymers being used in their place. The polymers tend to have dye fixing groups grafted along their molecular chain, to which the dyes are attracted. The bottom layer binds the ink vehicle and controls its dispersion into the paper bulk, so as to avoid cockle and curl. Similar problems with dye migration and bonding interactions can occur as with matt coated papers. Pigments can create other problems: for example, poor surface fixing can lead to poor rub resistance and low aggregation of pigment particles, leading to poor color strength. Too strong an attraction when the ink hits the paper can lead to poor orientation of pigment particles, and, in extreme cases, bronzing can occur. This can change over time, with the pigment particles changing to a more energetically favorable orientation with the polymer dye fixing groups.

Swellable coated papers are another type of glossy coated paper. Current market trends indicate that this type is mostly used on a polyethylene extruded photo film-type base, and so forms a different product. The "swellable" term comes from the ability of the polymer (usually PVOH or gelatin) to increase in size when absorbing the ink vehicle. After the ink vehicle enters the coating it is dispersed throughout the layer and the coating eventually shrinks to its original size. The speed at which this occurs depends on the particular ink system and the constituents. Therefore dot movement and resultant color change can occur, though this is less of a problem with new formulations. Dye migration and pigment reorientation within the layer can also occur during the return to size.

The principal layer of microporous papers is a coating of nanometer-size pores usually formed from the arrangement of silica in a high pigment-to-binder ratio. The pores enable the ink vehicle to be very quickly removed and dispersed through the paper. At the surface, dyes and pigments are held in a similar manner to that of the matt coated papers but the pore structure can lead to colorant movement. Depending on the size of pore, the dye molecule can travel into the coating, but is not actually chemically fixed within the pore. The pore will act as a capillary and the dye molecule can travel back up to the surface. The rate of travel will depend on the dye type, and hence there can be color changes over time.

Some pigments will do the same but due to their larger size tend not to enter the pores.

Note that there are a wide variety of coated media commercially available and only very general trends have been described above. This is especially true for combinations of coating types to allow a wider range of inks to be used.

22.2 Dye-Based and Pigment-Based Inks

Inkjet printers use either dye- or pigment-based inks. Dye-based inks tend to show lower light stability compared to pigment-based inks. Pigments tend to produce smaller color gamuts, though recent advances have increased the gamut producible with a pigment inkset. It is not possible to obtain good results using pigment-based inks with swellable media.

22.3 Trends by Paper and Ink Type

An investigation into this issue has been carried out by London College of Communication and Felix Schoeller GmbH on each of the main types of paper with dye and pigment inksets. CMYK primaries and their overprints were printed at 95%, 50%, and 10% tints and measured with a GretagMacbeth SpectroEye immediately after printing and then periodically over four days (swellable, matt, and gloss coated) and seven days (microporous). The environmental conditions were a constant 22 °C and 50% RH. Table 22.1 lists the average CIELAB color differences between the first measurement and the final measurement, together with color difference components ΔL^*, ΔC^*, and ΔH^*.

The following observations can be made:

1. There is a color shift for both inksets on all the media types.
2. Comparing the two inksets, the dye-based set has the higher color shifts with corresponding shifts in chroma and hue.
3. In all cases the prints get lighter with time while chroma falls.
4. The biggest lightness shifts occur with the microporous media for both ink types.
5. Color changes continued throughout the period of study, with no indication that a stable state had been reached.

If we were to rank the paper and ink combinations then the sequence would look like this (most stable first):

1. Matt coated + pigment ink
2. Gloss coated + pigment ink

Table 22.1 Average color differences for different media and ink types

	Dye-based ink		
Paper type	CIELAB ΔE_{ab}^*	CIELAB ΔL^*	CIELAB ΔC^*	CIELAB ΔH^*
Matt coated	1.23	0.57	0.94	0.55
Gloss coated	1.80	0.78	1.32	0.94
Microporous	1.90	0.85	1.51	0.78
Swellable	1.23	0.61	0.88	0.61
	Pigment-based ink		
Paper type	CIELAB ΔE_{ab}^*	CIELAB ΔL^*	CIELAB ΔC^*	CIELAB ΔH^*
Matt coated	0.87	0.68	0.53	0.12
Gloss coated	1.09	0.74	0.79	0.13
Microporous	1.22	0.81	0.80	0.44

3. Microporous + pigment ink
4. Matt coated + dye-based ink
5. Swellable + dye-based ink
6. Gloss coated + dye-based ink
7. Microporous + dye-based ink.

Therefore the use of pigment inks is to be recommended for stability of print, which is unsurprising given the inherent properties of pigments, including their inertness and particle size.

The findings given here are a summary of results, based on average measurements of the primary and secondary colors. The color shifts would also probably increase in magnitude at higher temperatures and humidity levels.

Offset litho and electrostatic printing processes were also tested using the same methodology by Helwan University, Cairo, and the results showed color shifts that can be regarded as not significant for most applications.

23

Viewing Conditions

The appearance of a color is significantly influenced by the illumination under which it is viewed. Perhaps the most important factors are the intensity and the spectral power distribution, or SPD (the relative amount of energy at each wavelength), of the illumination source.

Changing the SPD of the illumination alters the radiance reflected from a surface, since more energy will be reflected at those wavelengths that correspond to the highest relative power in the illumination. Although the human visual system has an outstanding ability to preserve the approximate appearance of a stimulus as the SPD of the illumination source changes, the retinal and cognitive mechanisms do not completely achieve color constancy. Moreover, in color management the goal is to produce a metameric match in which the required tristimulus values are defined but not the relative spectral power required to achieve this colorimetry. As a result a metameric match achieved under one illumination may fail under a different illumination.

Traditionally in graphic arts the colorants used in photographic media and printing inks had spectral reflectances that were very similar and so transparencies and prints matched quite consistently even when the viewing illumination was changed. Modern colorants (as used for example in dyes and toners in digital printing) often have quite different spectral reflectances from these traditional media, and can be particularly prone to mismatches arising from changes in viewing illumination.

In addition to the effect of the relative spectral power of the illumination in the visible region, the level of UV radiation in the illumination source will strongly affect the appearance of any materials that fluoresce.

In real viewing conditions there is typically a mix of some or all of incandescent, fluorescent, LED, and daylight illumination. The relative amounts of these may vary according to which lamps are illuminated at a particular time, the contribution of natural daylight through its intensity and the elevation of the sun, and any shading provided by window blinds, drapes, or curtains.

Since the appearance of a stimulus is likely to vary with the type of illumination, standardization of viewing conditions is essential in order to provide an agreed basis for the communication of color appearance and the assessment of color matches. It is important to note here that the illuminant used as a standard may not correspond to that of the actual illumination source in the end-use viewing condition, but a well-defined viewing condition is nevertheless essential as a reference for the purposes of data exchange. If the actual end-use viewing condition is known, then this can be used as the reference condition, but it is rare for the end-use viewing condition to be defined as unambiguously as is necessary.

The basic properties which may be used to define a viewing condition are the chromaticity and intensity of the illumination source, the reflectance of the background, and (where optically brightened substrates are viewed), the relative UV content of the source. A more complete specification is provided if the SPD of the illumination source, the relative luminance of the surround, and the chromaticity of the adopted white point are also defined.

In a specification of standard viewing conditions for reflective copy it is usually assumed that the adopted white is a perfect diffuse reflector, which will thus have the same chromaticity as the illumination source. For emissive and projected displays it is common to assume that the observer is completely adapted to the display white point and hence the chromaticity of the display white is taken as the adopted white.

Many industries involved in the manufacture of color products, such as paper, paints, and textiles, have agreed on standardization of CIE illuminant D65, which has a correlated color temperature of 6500 K, for measurement and viewing. D65 corresponds to average north-sky daylight. CIE daylight illuminant D50 (corresponding to a correlated color temperature of 5000 K and noon-sky daylight) is used in graphic arts, largely because it is closer to the chromaticity of indoor illumination and to the white point used in daylight photography.

23.1 PCS Viewing Condition

In situations where the source or destination viewing condition is not D50, the PCS-side values are chromatically adapted to D50. The ICC specification requires that in such cases the matrix used to chromatically adapt the adopted white point to D50 is specified in the profile in a chromaticAdaptationTag ("chad" tag), and if desired the CMM can use this to convert an image data encoding to the chromaticity of the actual source or destination viewing condition. Hence the viewing condition defined for the ICC perceptual PCS should be considered as part of the specification of a reference interchange encoding, not a requirement to actually use D50 in the color management workflow.

Equally, it may be desirable to evaluate proofs and final reproductions under the end-use viewing condition. This is not precluded by the specification of a reference viewing condition for the PCS, which is intended to provide a reference condition for the communication of appearance rather than a simulation of actual end-use viewing conditions.

This may seem to be an overcomplex solution in some situations, such as where a D65 display encoding is converted to a print encoding to be viewed under D65 illumination. In this case chromatic adaptation to D50 appears to be redundant. However, to achieve interoperability it is preferable to have a single reference viewing condition, with a well-defined procedure for transforming data between the reference viewing and all actual viewing conditions. The choice of D50 for the PCS reference viewing condition also means that it matches the actual viewing condition most commonly used in graphic arts.

If a source or destination profile is defined for a viewing condition that is not D50, profile generators can include a viewingCondDescTag which provides a textual description of the actual viewing conditions, and a viewingConditionsTag specifying the parameters of the actual viewing condition. The viewingConditionsTag enables the XYZ of the illuminant and surround to be stored in the profile as unnormalized CIE XYZ values, in which Y is in units of candelas per square meter and hence also implies the illuminance and surround relative luminance. The viewingCondDescTag can be used to distinguish between profiles generated for different viewing environments and to select one appropriate for the intended use.

23.2 Viewing Conditions and Rendering Intents

In versions of the ICC specification prior to v4, a single PCS and associated reference medium viewing environment were specified. The v4 specification introduced a distinction between the PCS used for colorimetric and perceptual rendering intents. The colorimetric PCS is now wholly measurement based, and as a result is no longer associated with a viewing condition. The perceptual PCS is now defined for a physically realizable medium with specified maximum and minimum luminances in an ISO 3664:2009 P2 viewing condition. This lower level (500 lux) is chosen for the ICC PCS since it is more typical of end-use viewing environments in the home and office than the higher ISO 3664:2009 P1 (2000 lux) level used in viewing booths for critical comparison of prints. It also corresponds to an adopted white luminance that is practically realizable on a color display in a home or office environment.

Since the colorimetric PCS is measurement based, inversion of the matrix stored in the chromaticAdaptationTag will produce values corresponding to the original medium colorimetry under the illuminant used to compute the original medium XYZ values. However, the PCS values stored for the perceptual intent will have been the result of a color rendering operation adjusting for factors such as dynamic range and gamut mapping, adaptation for differences between the PCS and end-use viewing conditions, and any further color adjustments applied to generate a preferred rendering. As a result the chromaticAdaptationTag is unlikely to produce either the original colorimetry or the optimal colorimetry (with preference adjustments) for the source viewing condition when inverted and applied to the PCS values for the perceptual intent.

23.3 Viewing Conditions for Prints, Transparencies, and Displays

Viewing conditions for graphic arts media are specified in ISO 3664:2009. This essentially specifies a D50 illuminant for color transparencies and prints, together with appropriate intensity levels and tolerances.

Reflection print viewing environments conforming with condition P1 should have an illuminance of 2000 lux. This produces an adopted white luminance of $636.6 \, \text{cd/m}^2$ for a perfect reflecting diffuser (since a diffuse reflector radiates $1/\pi$ of the incident flux).

Transparency illuminators conforming to condition ISO 3664:2009 T1 should have a luminance of $1250 \, \text{cd/m}^2$. When covered by a transparency whose base film has an assumed transmittance of 50%, the white point luminance is $625 \, \text{cd/m}^2$ and is sufficiently close to the adopted white luminance of the reflection print in the P1 condition for the user to have the same adaptation state when viewing transparency and print side by side.

Reflection print viewing environments conforming to condition P2 should have an illuminance of 500 lux, producing an adopted white luminance of $159.2 \, \text{cd/m}^2$ for a perfect reflecting diffuser.

Extraneous light and colored objects in the field of view should be avoided when performing assessments in a standard viewing environment.

Displays used for the appraisal of color images should have a white point chromaticity which approximates that of D65 and has a luminance of at least $80 \, \text{cd/m}^2$. When the display is used for direct comparison between soft copy images and prints viewed under a P2 condition, it is preferable for the user to have a single adopted white point and hence the display white point should be closer to the chromaticity of D50 and should have a luminance level of at least $160 \, \text{cd/m}^2$.

Ambient illumination in the display environment should be relatively low, so that the surround luminance is one-quarter or less of the luminance of the display white point. The correlated color temperature of the ambient illumination should be less than or equal to that of the display white point. The background against which images are displayed should have no more than 20% of the display white point luminance, and should ideally be 3% of the white point luminance. As with reflection print and transparency viewing, veiling glare and colored objects in the field of view should be avoided.

For substrates which are not completely opaque, the sample backing will have an effect on the color appearance and should be consistent with that used in practice. For measurement purposes the ICC recommends a white sample backing and for consistency this should also be applied to viewing.

23.4 Other Standard Viewing Conditions

There are many circumstances when the standard viewing conditions for prints, transparencies, and displays defined in ISO 3664 are not relevant, particularly in the case of source images encoded in reference and interchange color encodings such as the ISO 22028 and IEC 61966-2 series. The viewing conditions associated with these encodings will differ from that of the ICC PCS, and hence require chromatic adaptation (and possibly an appearance model transform, if illuminance and surround relative luminance are different) to the ICC PCS. Details of these viewing conditions can be found in the relevant standards.

23.5 Viewing Conditions and Measurement

The aim of color measurement is to provide a metric which correlates with visual perception. This implies that the geometry and spectral responsivity of the measurement instrument should ideally simulate those of the human visual system. For reflective samples, it also implies that the sample should be illuminated with a source having the same SPD as is used by the observer when viewing the sample. For this reason, ISO TC 130 and ISO TC 42 have collaborated in joint working groups to produce revisions to ISO 13655 and ISO 3664 that harmonize the SPD of both instrument and standard viewing condition.

ISO 3664:2009 specifies that reflective samples shall be judged in a viewing environment having a source corresponding to CIE daylight illuminant D50. ISO 13655:2010 specifies four source SPDs for measurement instruments, of which M1 is recommended for measurement of graphic arts samples. In both cases the SPD of D50 is required to include the spectral power of D50 that lies outside the visible range and in the UV, which is essential to ensure a correspondence between measurement and appearance when fluorescent whitening agents are present in the sample – as is the case for almost all commercial printing papers.

Where the source used in the end-use viewing condition includes little or no UV, the ICC recommends that measurements are made with the ISO 13655 M2 measurement condition, which excludes UV from the source (see Chapters 20 and 21 for more details). This will lead to a better prediction of the appearance in the end-use viewing condition. It does have the effect that the UV component in the standard viewing condition may give rise to a mismatch between proof and print, where the proof has been made on a substrate with a different amount of FWA than that of the final print. In this situation it may be desirable to temporarily mask the UV component from the viewing booth source to evaluate the match in the absence of fluorescent excitation, although users should be aware that this viewing condition does not conform to ISO 3664.

For display measurement, ISO 13655 specifies that XYZ values should be computed from the spectral power of the display emission (without a standard illuminant, since the display is self-luminous), and may be normalized by dividing by $100/Y_w$, where Y_w is the Y tristimulus value of the adopted white. ICC color management assumes that the user is completely adapted to the display white point: the display colorimetry is normalized to the display white and chromatically adapted to the PCS D50 white point. For most applications this will produce an optimal conversion between a source image on a display and a reproduction on another medium with a different viewing condition. The chromaticAdaptationTag matrix can be used to recover the original colorimetry in the source viewing environment if required.

23.6 Assessment of Viewing Conditions

Aim values for reference viewing conditions are defined in ISO 3664. A particular realization of this reference viewing condition can be assessed in terms of its ability to meet the SPD, chromaticity, and intensity of illumination. Details of such assessment are given in ISO 3664:

- Tolerances for the luminance or illuminance are approximately 25% of the aim value, with departures from uniformity no greater than 25% between center and edge.
- The chromaticity of the illumination is required to be within a radius of 0.005 from the aim values specified for D50.
- The SPD of the illumination is evaluated with respect to the CIE daylight illuminant D50 by means of a color rendering index (CRI) of at least 90 and metamerism index in the UV (MI_{UV}) of less than 1.5.

Computational procedures for calculating these parameters are given in ISO 3664.

For the practical evaluation of a viewing condition, it is essential to use a telespectroradiometer with a narrow spectral bandpass (<5 nm) calibrated to a standard source traceable to a national standardizing laboratory. Illuminance and SPD can be measured directly if a cosine-correcting diffuser is fitted, or alternatively luminance and SPD can be measured from a sample of known reflectance (such as a calibrated reference material), whereby the measured values are divided by the sample reflectance to obtain the corresponding values for the incident illumination. Where only illuminance is being assessed, a simple photometer (preferably traceable to a national standardizing laboratory) is sufficient.

23.7 Viewing Condition and Color Appearance

The viewing condition under which a color stimulus is viewed strongly influences its appearance. Colorimetric coordinates can be computed for a given viewing condition, but to predict the appearance of the same stimulus under a different viewing condition – or to predict the colorimetry required to match the original stimulus in a different viewing condition – a model of color appearance is required. Currently the CIECAM02 model is recommended by the CIE for this purpose.

Although the details of calculating appearance coordinates using models such as CIECAM02 are outside the scope of this chapter (readers are referred to the literature describing these models), a brief summary is given here of the effect of the viewing condition on color appearance and how these effects should be interpreted or applied within an ICC color management workflow.

Where the source and destination image colorimetry are defined for the same viewing condition, the appearance model should predict the same XYZ coordinates for both source and destination conditions, and is therefore not required. In situations where the only change in the viewing conditions between source and destination is a change in the chromaticity of the adopted white point, a chromatic adaptation model is sufficient to predict the change in XYZ coordinates required to match the original under the source conditions to the reproduction under the destination conditions. Only where other differences between source and destination viewing conditions exist is an appearance model required in order to predict the final appearance.

For the purpose of modeling color appearance, a number of terms can be defined. The stimulus forming the focal color perception is assumed to extend to 2° of angular subtense. The *background* is the region subtending approximately 10° beyond this stimulus. The *adapting field* includes everything outside the background, while the *surround* is a categorical representation of the ambient illumination in comparison to the image white point luminance. Surround categories in CIECAM02 are average, dim, and dark.

The *adapted white point* is the internal human visual system white point for a given set of viewing conditions, while the *adopted white point* is the white point actually used in the calculation of appearance coordinates and white point normalized colorimetry.

In CIECAM02, the adapting luminance L_A is the luminance of the adapting field. In most cases a "Grey World" assumption is made and the adapting luminance calculated as one-fifth of the luminance of the adopted white point.

The background relative luminance parameter Y_b is calculated by dividing the luminance of the background by the luminance of the adopted white point.

Color appearance models were derived primarily for simple color stimuli, but are generally applicable to complex images. The main issue to take into consideration is how the image background is to be defined. Some studies have found that taking the mean luminance of a complex image provides an adequate description of the background effect on a given pixel. Alternatively a "Grey World" assumption is sometimes made and the Y_b parameter is set based on a neutral gray background with a reflectance of 20%.

If a profile includes a viewingConditionsTag, the L_A parameter is found by dividing the Y tristimulus value of the illuminant by 5. The surround category is found by comparing the Y tristimulus values of the surround and illuminant: if the ratio of surround to illuminant is 20% or above, the average surround category is chosen; if the ratio is below 20% the dim surround category is appropriate; and finally if the ratio is close to 0 the dark surround category should be chosen.

ISO 3664 P1 and P2 conditions (and hence the ICC perceptual PCS and most print-centric tasks such as print and proof viewing) imply a CIECAM02 "average" surround condition, while display-centric calibrated RGB encodings are typically based on a surround relative luminance which corresponds to a "dim" category. For other media, such as projection displays, a "dim" surround may be applicable.

The XYZ of the illuminant stored in the viewingConditionsTag also provides the data required to perform chromatic adaptation from the PCS D50 illuminant to other illuminants as required, either by using a chromatic adaptation transform such as the CAT02 or Bradford models, or in a color appearance transform in which chromatic adaptation is an element in the computational procedure.

A input profile can also include a colorimetric intent image state tag ("ciis"), which specifies how the data for the colorimetric intent stored in the profile should be interpreted. Four signatures are currently supported:

- scene colorimetry estimate "scoe"
- scene appearance estimate "sape"
- focal plane colorimetry estimate "fpce"
- reflection hard copy original colorimetry "rhoc."

For the first three of these signatures, the adopted white is normalized to 1.0 as with reflection print and display colorimetry, but the mediaWhitePointTag Y tristimulus value is relative to the adopted white Y value and can be larger than 1.0. This allows the calculation of appearance effects which depend on viewing condition parameters such as the luminance of the adapting field, and also makes it possible to communicate the appearance of a scene containing specular highlights with luminances greater than that of the diffuse white point of the destination media.

Examples of typical corrections to media-relative colorimetry that are required in a color management workflow are the adjustment of brightness and colorfulness to compensate for the effect on tonal reproduction of a dark background or an adapting field whose illuminance is significantly lower or higher than the PCS perceptual intent viewing condition (such as print appraisal under typical office lighting with 200–400 lux illuminance, or proof viewing in an ISO 3664 P1 condition with 2000 lux). A dark background, such as a transparency rebate, gives rise to an impression of a brighter image, especially in shadow areas. The contrast and colorfulness of a reflective print will increase with increasing incident illumination, while that of a display will fall. Both these effects can be modeled by CIECAM02 with reasonable success.

Part Five

Profile Construction and Evaluation

24

Overview of ICC Profile Construction

An ICC profile contains the color processing elements required to transform data between the profile connection space and the data encoding of the profile. The particular elements to be included are specified for each profile class in Annex G of the ICC specification. The profile will also include additional data to help the CMM to interpret the encoded transform correctly, and to help the user or workflow system to select the profile for a given conversion (or to select particular color processing elements with the profile).

Full information about the data to include in a profile is given in the specification, which should be the primary reference for anyone building ICC profiles. Detailed guidance on particular aspects of profile construction and use is provided in other chapters; in this chapter the goal is to provide an overview of the process of profile creation and where appropriate to point to other sources of information. It should be noted that references to section numbers in the ICC profile specification refer to numbering in Version 4.3 of the specification (ISO 15076-1, revised 2008–2009, also known as ICC.1:2010).

All valid profiles require a 128-byte header, a tag table identifying the tags present in the profile, a description string, the numerical values of the media white, and the color processing elements and other tags as defined for the profile class by the specification. Profiles may optionally include other valid tags, together with private tags not defined in the specification but registered in the ICC Signature Registry. In general the ICC discourages the use of such private tags as they may limit the interoperability of the profile and lead to inconsistent results, since CMMs may not know how the private tag is to be interpreted.

The ICC profile format is a binary format which contains all the information required to transform color data between the data encodings represented by the profile. For most profile classes these encodings will consist a data encoding representing a device or color space of some kind on one side, and the ICC PCS on the other. This allows a CMM to connect source and destination profiles for a transform unambiguously, regardless of the applications or operating system used. The tag structure of the profile format provides a baseline functionality, which can be extended in well-defined ways as needed.

24.1 Why a Binary Format?

A binary format of this kind is of course not the only way of encoding a color transform. Possible alternatives include:

1. A procedural definition of the transform, in which the "profile" includes the code as well as the data to apply the transform.
2. A text file which provides the transform data only.
3. An XML-based format in which the color processing elements and other elements of the "profile" are specified in a schema.

Each of these has applications in particular workflows. A procedural definition may ensure that the transform cannot be misinterpreted, although it has considerably less flexibility and platform independence, and may require the use of proprietary intellectual property in the procedure used. A simple text file may be useful in workflows where the procedural implementation is well defined and a simple method of encoding new processing elements is desired. And an XML-based format can take advantage of the properties of XML, which provides a mechanism for developing structured content with well-defined semantics. Text-based and XML-based formats also have the advantage that they can, if desired, be human readable and editable in a simple text editor.

Internal ICC projects have shown that it is feasible to convert programmatically between the ICC binary profile format and other profile formats such as XML or text files. The current ICC format enables a high level of interoperability across a very wide range of application programs and operating systems. At the same time it contains sufficient flexibility to support new applications (as can be seen for example in the chapters on digital photography and on the use of multi-processing elements), and to enable dynamic and programmable run-time color transforms where needed. In the ICC architecture there is a well-defined baseline interpretation of the color data, while at the same time developers for specific applications and workflows are free to implement extended functionality in interpreting and applying the transform by adding features to the CMM.

24.2 Writing Profiles

When a profile is generated, the required elements including text strings and numerical values are written at specified locations in a file, which is given the extension ".icc" or ".icm."

Powerful applications that will generate a wide range of profile classes, and guide the user through the process of obtaining the measurement data needed, are available from ICC members and listed on the Profiling Tools page on the ICC web site.

Profiles can also be written using calls to library functions. Widely used examples are the C++ libraries SampleICC, lcms, and Argyll, and the routines in Mathworks' MATLAB Image Processing Toolbox. Use of these functions requires some knowledge of programming and (in the case of the C++ libraries) the use of compilers, a knowledge of the data types and encodings used in the specification, knowledge of the color processing models used in the profile format specification and an understanding of how to characterize a device or data encoding in order to produce the values required for the color processing elements of the profile.

Profiles can sometimes be created by inserting the required data into an existing profile. However, the byte offsets of the profile must be respected and the new elements must be consistent with all the other data in the profile, such as the media white point. Arbitrarily changing some of the data in a transform is likely to have undesirable consequences, such as introducing discontinuities or unwanted color shifts in images converted with the new profile, and will also invalidate the profileID value in the header, if present. As a result this approach only works in special cases where a small change in a header attribute or a component of the color processing model is altered but other elements are essentially unchanged; in general the method is not recommended. It also runs the risk of infringing the intellectual property of the creator of the original profile. Nevertheless, beginning with a known good profile can be a good way of testing out different elements of a profile and the encoding methods required.

24.3 Data Encodings

The ICC profile defines a variety of different data encodings, which permit all the values required in a profile to be encoded efficiently without excessive proliferation of data types.

24.4 Unsigned Integer Types

Unsigned integer encodings of 1, 2, 4, and 8 bytes are defined as uint8Number, uint16Number, uint32Number, and uint64Number respectively. For a domain [0,1] to be encoded as a uint8Number, decimal values are multiplied by 255. For a domain [0,1] to be encoded as a uint16Number, decimal values are multiplied by 65 535.

The uint8Number data type is used in all the LUT types to specify the number of channels and grid points. In a lut8Type, the uint8Number is additionally used to define the input and output tables. Uint8Number encodings are also used in the colorantOrderType encoding to give the number of each colorant in the sequence.

The uint16Number type is used in 16-bit LUT-type encodings to provide the color LUT (CLUT) grid points and also (in the lut16Type encoding) to specify the number of entries in the input tables and output tables, together with the values of tables themselves.

The uint16Number type is also employed in other tag types in the specification, including:

- Entries in the dateTimeNumber type
- Normalized device value in a response16Number
- Number of device channels in a chromaticityType
- PCS values of the colorants in a colorantTableType
- PCS values in a namedColor2Type
- Curve values in a curveType
- Language code and country code in a multiLocalizedUnicodeType
- Function type in a parametricCurveType.

Additionally, uint16Numbers are used to define the number of channels and count of measurement types in a responseCurveSet16Type, where they are encoded as part of a response16Number.

The uint32Number type is used to encode the following:

- Profile rendering intent in the profile header
- Count of tags in the profile tag table
- Count of colorants in the colorantOrderType and colorantTableType encodings
- Offsets in lutAToBType and lutBToAType encodings
- Count of entries in curveType encoding
- String length and string offset in the multiLocalizedUnicodeType
- Count of named colors and count of device coordinates for each named color in namedColor2Type encoding.

Additionally, uint32Numbers are used to define the offsets and counts of measurements per channel in a responseCurveSet16Type. Full details of numeric types used in ICC profiles are given in Section 4.2.1 of the ICC specification, "Basic number types."
Currently there are no tags that make use of uint64Numbers.

24.5 Fixed Number Types

The specification defines four fixed number types. These include the two 16-bit types u8Fixed8Number and u1Fixed15Number, and the two 32-bit types u16Fixed16Number and s15Fixed16Number. The fixed number types have an implied decimal point, so that the number consists of an integer part and a fractional part. Since fixed numbers have limited bits of precision, many decimal numbers can only be encoded approximately.

The u8Fixed8Number is a fixed unsigned 2-byte quantity with 8 fractional bits, used to encode exponents in the curveType encoding. To convert from decimal to u8Fixed8Number, multiply by 256.0 and convert to a 16-bit unsigned integer

For example, for a γ value of 2.2:

$$256.0\gamma = 563.2$$

$$\text{u8Fixed8Number} = \text{round}(563.2) = 563 \ (\text{hex } 0233\text{h}).$$

Note that this is not exactly 2.2, but the closest u8Fixed8Number equivalent corresponding to $\gamma \approx 2.199\,218\,75$.

The u1Fixed15Number is a fixed unsigned 2-byte quantity with 15 fractional bits, used to encode PCSXYZ values. To convert from decimal to u1Fixed15Number, multiply by 32 768.0 and convert to a 16–bit unsigned integer.

For example, for PCSXYZ $Z = 1.08$:

$$32\,768.0 \ \text{PCSXYZ} \ Z = 35\,389.44$$

$$\text{u1Fixed15Number} = \text{round}(35\,389.44) = 35\,389 \ (\text{hex } 00008\text{A3Dh})$$

corresponding to PCSXYZ $Z \approx 1.079\,986\,572$.

The u16Fixed16Number is a fixed unsigned 4-byte quantity with 16 fractional bits, used to encode x, y chromaticity values and measurement flare. To convert from decimal to u16Fixed16Number, multiply by 65 536.0 and convert to a 16-bit unsigned integer.

For example, for a chromaticity y of 0.3:

$$65\,536.0y = 19\,660.8$$

$$\text{u16Fixed16Number} = \text{round}(19\,660.8) = 19\,661 \text{ (hex 00004DF9h)}$$

corresponding to $y \approx 0.299\,987\,793$.

The s15Fixed16Number is a fixed signed 4-byte quantity with 16 fractional bits, used in matrix coefficients and parametric curve parameters. Positive decimal values are converted to the s15Fixed16Number encoding in the same way as for the u16Fixed16Number, while negative values are encoded using two's complement, which means that the result is added to 2^{32} or 4 292 967 296.

Hence to convert x from decimal to s15Fixed16Number, multiply by 65 536.0 and convert to a 32-bit signed integer.

For example, for a matrix coefficient $e = -1.82$:

$$65\,536.0e = -119\,275.52$$

$$\text{s15Fixed16Number} = \text{round}(-119\,275.52) = -119\,276$$

or two's complement value of 4 294 848 020 (hex FFFE2E13h) corresponding to $e \approx -1.820\,007\,324$.

The specification also provides for arrays of the above unsigned integer and 32-bit fixed data types, defined as uInt8ArrayType, uInt16ArrayType, uInt32ArrayType, uInt64ArrayType, u16Fixed16ArrayType, and s15Fixed16ArrayType. The derivative types dateTimeNumber, response16Number, and XYZNumber are also based on the data types described above.

24.6 Floating Point Encoding

The float32Number specifies a 32-bit floating point number, defined as in IEEE 754. Floating point numbers are used in the multiProcessElementsType and its BToDx and DToBx tags.

24.7 Profile Header

The header of an ICC profile is the first 128 bytes, populated as described in Section 6.2 of the specification. The header contains essential metadata which identifies the profile and defines the intended uses of the profile. Much of this information is invariant across a particular class of profiles, the only data elements which are required to be profile specific being the profile size, creation date, and (where present) profileID.

The header uses 4-byte signatures and binary flags for fields which must match an item in an enumerated list, together with numerical values for the profile size, PCS illuminant, and profileID. This approach eliminates possible ambiguity and helps in making the profile robust since CMMs should have little difficulty in interpreting the data provided in the header.

24.7.1 Header Enumerated Values

Signatures for the profile class, data color space, profile connection space, and primary platform should be selected as appropriate from the enumerated lists provided in Tables 14–16 of the ICC specification. For all profile classes except DeviceLink, the PCS has the signature "XYZ" or "Lab." A DeviceLink profile connects two or more device encodings, and the PCS is the data color space of the last device encoding in the sequence.

Signatures for the CMM type, device manufacturer, and device model are provided in the ICC Signature Registry. If a profile creator wishes to register a new signature for manufacturer or model, this can be done online. To register a new CMM, the profile creator should contact the ICC Technical Secretary. It should be noted that the CMM type identified in the profile header only indicates the preferred CMM of the profile creator, and the user or application is free to select a different CMM at run-time.

The rendering intent field identifies the intent preferred by the profile creator, and like the CMM type the actual rendering intent used when converting between profiles is subject to choices made by the user and the application. Values for this field can be 0, 1, 2 or 3, corresponding to perceptual, media-relative, saturation, and ICC-absolute colorimetric intents, encoded as a uint32Number.

In the case of a DeviceLink profile, the rendering intent specifies the intent actually used when combining two or more profiles to make the link profile.

24.7.2 Header Numeric Values

The profile size is the number of bytes of the entire file, including the 128-byte header, the tag table, and the tagged element data. This number should be divisible by 4, since all data is required to be padded to 4-byte boundaries. The size in bytes is encoded as a uint32Number.

The creation date is encoded as a dateTimeNumber, a 12-byte value consisting of 2-byte fields for each of the year, month, day, hour, minute, and second when the profile was first created.

The PCS illuminant is required to have the values $X = 0.9642$, $Y = 1.0$ and $Z = 0.8249$. These values are encoded as an XYZNumber (i.e., [63 190, 65 536, 54 061] or [0F6D6h, 10000h, 0D32Dh]).

The profileID field is intended to hold a 16-byte hash value associated with the profile content. It is determined by first setting the profile flags, rendering intent, and profileID fields to zero, and then computing the MD5 hash value for the file. If the profileID is not computed, the field should be set to hexadecimal zero (00h).

24.7.3 Flags and Device Attributes Fields

The profile flags field has 2 bits set by the profile creator, the first indicating whether the profile is intended to be embedded in another file, and the second indicating whether, if embedded, the profile can be used independently. Thus the flag in bit position 1 can only be set to 1 if the flag in bit position 0 is also set to 1.

The device attributes field consists of four 1-bit values indicating basic media attributes, 28 reserved bits (set to zero), and 32 vendor-defined bits.

24.7.4 Other Header Fields

The profile version number specifies the version of the profile format with which the profile conforms. The current version, ICC.1:2010, is encoded as 04300000h. It is recommended that profile creators use the latest available version of the profile format, or if creating profiles according to the v2 specification, the ICC.1:2001–04 specification should be used, and the profile version encoded as 02400000h.

The profile signature field should correspond to the signature "acsp," encoded as 61637370h. Note that older versions of the specification incorrectly specified the hex encoding as 61637379h.

24.8 Tag Table

The tag table follows immediately after the profile header. It consists of a count of the number of tags in the profile, encoded as a uint32Number, followed by a list of the tags. For each tag, the 4-byte tag signature is given, followed by a uint32Number specifying the offset from the beginning of the file to the tag element data, and a uint32Number specifying the size of the tag element data in bytes. The tags can be listed in any order, and where the same element data is to be reused for two or more tags, the offsets can point to the same element data. All tag data elements should start on a 4-byte boundary, with unused bytes being padded with binary zeros.

24.9 Tag Elements

A complete listing of ICC tags is shown in Table 24.1, where the tag elements which are required to be included for each profile class are identified. Other tags which are defined in the ICC specification can be included in a profile, but as profile consumers are not required to interpret them, the results from applying them are undefined and will be implementation dependent. To maintain interoperability, the ICC does not recommend incorporating such additional tags where they affect the outcome of the color processing model defined by a profile, or where the profile cannot be correctly interpreted without them.

Vendors may also include private tags in a profile, and the signatures for such private tags should be registered in the ICC Signature Registry. Their usage will be implementation dependent and so their inclusion in a profile is not generally recommended by the ICC.

While there is not space in this chapter to detail the mechanics of the construction of every profile class and the possible combination of tags they can contain, the construction of tag elements for three common types of profile are briefly summarized below. Note that while the steps described below are typically necessary when constructing a conforming ICC profile, they may not be sufficient to produce a good-quality profile for a particular application.

Table 24.1 Tags defined in the ICC 4.3 specification

Name	Description	Required in:
AToB0Tag	Multi-dimensional transform structure	LUT-based input display and output; DeviceLink; abstract
AToB1Tag	Multi-dimensional transform structure	LUT-based output
AToB2Tag	Multi-dimensional transform structure	LUT-based output
blueMatrixColumnTag	The third column in the matrix used in matrix/TRC transforms	Matrix-based input and display
blueTRCTag	Blue channel tone reproduction curve	Matrix-based input and display
BToA0Tag	Multi-dimensional transform structure	LUT-based display and output
BToA1Tag	Multi-dimensional transform structure	LUT-based output
BToA2Tag	Multi-dimensional transform structure	LUT-based output
BToD0Tag	Multi-dimensional transform structure supporting float32Number-encoded input range, output range, and transform	
BToD1Tag	Multi-dimensional transform structure supporting float32Number-encoded input range, output range, and transform	
BToD2Tag	Multi-dimensional transform structure supporting float32Number-encoded input range, output range, and transform	
BToD3Tag	Multi-dimensional transform structure supporting float32Number-encoded input range, output range, and transform	
calibrationDateTimeTag	Profile calibration date and time	
charTargetTag	Characterization target such as IT8/7.2	
chromaticAdaptationTag	Converts an nCIEXYZ color relative to the actual adopted white to the nCIEXYZ color relative to the PCS adopted white	All except DeviceLink. Required only if the chromaticity of the actual adopted white is different from that of the PCS adopted white
chromaticityTag	Chromaticity values for phosphor or colorant primaries	
colorantOrderTag	Identifies the laydown order of colorants	

Tag	Description	Profile types
colorantTableTag	Identifies the colorants used in the profile. Required for N-component-based output profiles and DeviceLink profiles only if the data color space field is xCLR (e.g., 3CLR)	LUT-based output. Required only if the data color space field is xCLR
colorantTableOutTag	Identifies the output colorants used in the profile, required only if the PCS field is xCLR (e.g., 3CLR)	DeviceLink required only if the PCS field is xCLR
colorimetricIntentImageStateTag	Image state of PCS colorimetry resulting from the use of the colorimetric intent transforms	All
copyrightTag	Profile copyright information	
DToB0Tag	Multi-dimensional transform structure supporting float32Number-encoded input range, output range, and transform	
DToB1Tag	Multi-dimensional transform structure supporting float32Number-encoded input range, output range, and transform	
DToB2Tag	Multi-dimensional transform structure supporting float32Number-encoded input range, output range, and transform	
DToB3Tag	Multi-dimensional transform structure supporting float32Number-encoded input range, output range, and transform	
deviceMfgDescTag	Displayable description of device manufacturer	
deviceModelDescTag	Displayable description of device model	
gamutTag	Out of gamut: 8-bit or 16-bit data	LUT-based output
grayTRCTag	Gray tone reproduction curve	Monochrome input, display, and output
greenMatrixColumnTag	The second column in the matrix used in matrix/TRC transforms	Matrix-based input and display
greenTRCTag	Green channel tone reproduction curve	Matrix-based input and display
luminanceTag	Absolute luminance for emissive device	
measurementTag	Alternative measurement specification information	
mediaWhitePointTag	nCIEXYZ of media white point	All except DeviceLink
namedColor2Tag	PCS and optional device representation for named colors	NamedColor
outputResponseTag	Description of the desired device response	
perceptualRenderingIntentGamutTag	Gamut adopted as reference medium for the perceptual rendering intent	

(continued)

Table 24.1 (*Continued*)

Name	Description	Required in:
preview0Tag	Preview transformation: 8-bit or 16-bit data	
preview1Tag	Preview transformation: 8-bit or 16-bit data	
preview2Tag	Preview transformation: 8-bit or 16-bit data	
profileDescriptionTag	Structure containing invariant and localizable versions of the profile name for displays	All
profileSequenceDescTag	Array of descriptions of a sequence of profiles used to generate a DeviceLink profile	DeviceLink
profileSequenceIdentifierTag	Structure containing information identifying the sequence of profiles used in generating a DeviceLink	
redMatrixColumnTag	The first column in the matrix used in matrix/TRC transforms	Matrix-based input and display
redTRCTag	Red channel tone reproduction curve	Matrix-based input and display
saturationRenderingIntentGamutTag	Gamut adopted as reference medium for the saturation rendering intent	
technologyTag	Device technology information such as LCD, CRT, dye sublimation, and so on	
viewingCondDescTag	Viewing condition description	
viewingConditionsTag	Viewing condition parameters	

24.10 Three-Component Matrix-Based Display Profiles

In addition to the profileDescriptionTag, mediaWhitePointTag, and copyrightTag required by all profiles, a three-component matrix-based profile must have tags for the XYZ values of the three primaries, the tone reproduction curve for each component, and a chromaticAdaptationTag. The color processing model for three-component matrix-based profiles is described in Annex F of the specification.

The XYZ values of the three primaries are encoded as an XYZNumber and (in a v4 profile) stored as a redMatrixColumnTag, greenMatrixColumnTag, and blueMatrixColumnTag, with signatures "rXYZ," "gXYZ," and "bXYZ." The tone reproduction curves (TRCs) are (in a v4 profile) encoded as a curveType or parametricCurveType, with signatures "rTRC," "gTRC," and "bTRC."

The values to be encoded are determined as follows. First, measurements are obtained of the three primaries, the peak white, and sufficient additional colors to define the tone reproduction curve. The color processing model for three-component matrix-based profiles assumes additivity of the three primaries, so this should be tested: if the sum of the X tristimulus values of the primaries approximates the X of the media white (and similarly for the Y and Z values), the device exhibits additivity. If the sums of the X, Y, and Z tristimulus values for the primaries do not approximate the media white X, Y, and Z, a three-component matrix-based profile will not give a very accurate model and an LUT-based profile should be considered instead.

The next step is to scale all the measurement data to be media-relative PCSXYZ:

$$X_{PCS} = \left[\frac{X_{D50}}{X_{mw}}\right] X_n$$

$$Y_{PCS} = \left[\frac{Y_{D50}}{Y_{mw}}\right] Y_n \qquad (24.1)$$

$$Z_{PCS} = \left[\frac{Z_{D50}}{Z_{mw}}\right] Z_n$$

where:

X_n, Y_n, Z_n are the measurement data relative to a perfect diffuser, computed using the D50 illuminant, and normalized so that $Y = 1$ for the perfect diffuser (this is referred to as nCIEXYZ in the ICC specification);

X_{mw}, Y_{mw}, Z_{mw} are the nCIEXYZ values of the media white point as specified in the mediaWhitePointTag of the profile;

X_{D50}, Y_{D50}, Z_{D50} are the nCIEXYZ values of the PCS white point (i.e., [0.9642, 1.0, 0.8249]).

The PCS side of the profile is required to be D50, so if the measurement data is not computed using the D50 illuminant it is necessary to apply chromatic adaptation to the data so that the nCIEXYZ values in Equation (24.1) are based on D50. The profile creator is free to use any chromatic adaptation transform to achieve this, the most common choice being a linearized version of the Bradford chromatic adaptation transform. This process results in a media white chromatically adapted to D50, which is stored in the mediaWhitePointTag as an XYZType.

The chromatic adaptation to D50 is given by:

$$
\begin{bmatrix} X_n \\ Y_n \\ Z_n \end{bmatrix} = \text{CAT} \begin{bmatrix} X_{\text{SRC}} \\ Y_{\text{SRC}} \\ Z_{\text{SRC}} \end{bmatrix}
\tag{24.2}
$$

where X_{SRC}, Y_{SRC}, Z_{SRC} represent the measured nCIEXYZ values in the actual device viewing condition, X_n, Y_n, Z_n represent the chromatically adapted nCIEXYZ values, and CAT is the chromatic adaptation transform.

If the chromatic adaptation transform used in computing D50 colorimetry has the form of a 3×3 matrix, this matrix is stored in the chromaticAdaptationTag (encoded as s15Fixed 16ArrayType). If the transform used does not have the form of a 3×3 matrix, the chromatic-AdaptationTag matrix can be calculated for an additive system as follows:

$$
\text{CAT} = \begin{bmatrix} Xr_n & Xg_n & Xb_n \\ Yr_n & Yg_n & Yb_n \\ Zr_n & Zg_n & Zb_n \end{bmatrix} \begin{bmatrix} Xr_{\text{SRC}} & Xg_{\text{SRC}} & Xb_{\text{SRC}} \\ Yr_{\text{SRC}} & Yg_{\text{SRC}} & Yb_{\text{SRC}} \\ Zr_{\text{SRC}} & Zg_{\text{SRC}} & Zb_{\text{SRC}} \end{bmatrix}^{-1}
\tag{24.3}
$$

where $Xr_{\text{SRC}}, Yr_{\text{SRC}}, Zr_{\text{SRC}}$ represent the measured nCIEXYZ values for the red colorant in the actual device viewing condition and Xr_n, Yr_n, Zr_n represent the chromatically adapted values for the red colorant; and similarly for the green and blue colorants.

Since the chromatic adaptation between device encoding and the PCS is incorporated in the profile, no further processing is required when transforming between source and destination encodings. The chromaticAdaptationTag is optionally used by a CMM, for example, to calculate the original XYZ values before chromatic adaptation was applied. More details of the chromaticAdaptationTag and its use are given in Annex E of the ICC specification.

In a three-component matrix-based display profile, it is assumed that the adopted white is the display peak white, and hence the D50 illuminant is stored in the mediaWhitePointTag. In a three-component matrix-based input profile, however, the normalization step is not performed and so the value stored is the measured white point of the media after chromatic adaptation. In the v2 specification the display profile requirements were ambiguous, and the changes made on introducing the v4 specification are outlined in Chapter 10.

24.11 Three-Component LUT-Based Input Profile

A three-component LUT-based input profile is only required to have an AToB0Tag in addition to the profileDescriptionTag, mediaWhitePointTag, and copyrightTag required by all profiles, together with a chromaticAdaptationTag if the input media measurements are not D50. The color processing model for three-component LUT-based input profiles is described in Annex F of the specification.

Details of constructing LUTs for ICC profiles are given in Chapter 28. For an input profile AToB0Tag, a profile creator would first obtain measurements which sample the input medium (using a test chart such as the ISO 12641 chart, also known as IT8.7/1 and IT8.7/2 for transparent and reflective media respectively). This test chart would be imaged by the scanner or camera for which an input profile is to be made, and for each color patch an RGB value determined (usually computed as an "average" of multiple pixels within the color patch).

Having acquired the colorimetric data and corresponding device data, the next step is to convert the data to media-relative colorimetry, as described above, and, if the measurements are not D50, to chromatically adapt the data to D50.

Next, the relationship between the two data sets (the RGB device values and the media-relative chromatically adapted colorimetric values) is characterized, using a suitable mathematical model. In the simplest profile structure the matrix and curve elements of the LUT tag are not used (or set to identity) and the CLUT converts directly from the device encoding to the PCS. A uniformly spaced input table sampling the entire device encoding is created, and the device model used to predict the PCS values for each entry in the input table.

Other elements in the LUT type can be used to improve the color processing model. A curve to be applied to the RGB values might be computed so that the values output from the curve are linear with respect to the PCSXYZ encoding. The matrix might then be used to perform a linear conversion between linearized RGB and XYZ primaries, so that the output of the matrix is optimized for the CLUT. Additionally, curves before and after the CLUT might be used to give greater weight to neutrals. More detail on LUT processing elements is given in Chapters 25 and 28.

After the matrix and curve elements are determined, the device model is generated to convert between the output of the elements which are processed prior to the CLUT (the A curve in a v4 lutAToBType, or the input table in a v2 lut8Type), and the input to the elements processed after the CLUT (the M curve, matrix, and B curve in a v4 profile, or the output table in a v2 profile). Again a uniformly spaced input table sampling the entire domain is generated, and the corresponding output values computed using the device model.

Finally, to encode the data for each element the values are normalized to the maximum of the data type. For example, if the data type is uint16Number, the data is scaled so that the maximum value is 65 535.

For the media-relative colorimetric intent, the steps above generate the color processing elements needed for the profile. However, in many cases additional adjustments are require to render from the input medium to the output-referred PCS. This might include compensation for differences in viewing conditions, preference adjustments performed to generate a more preferred output, and a gamut compression or expansion so that the PCS values which result from the device encoding span the Perceptual Reference Medium Gamut. Such adjustments to the colorimetric data are encoded as the perceptual rendering intent.

An input profile is not required to have additional rendering intents beyond the AToB0Tag, but if desired the full set of AtoBx and BToAx tags can be encoded. Further discussion of rendering intents can be found in Chapters 13, 12 and 25.

24.12 Four-Component LUT-Based Output Profile

A four-component LUT-based output profile has a similar structure to a three-component LUT-based input profile, except that the full set of AtoBx and BToAx tags must be included. A gamutTag is also required, to indicate which values in the PCS encoding are outside the effective gamut of the output encoding. The color processing model for four-component LUT-based output profiles is described in Annex F of the specification.

The AtoB0, AtoB1, andAtoB2 tags encode the perceptual, media-relative, and saturation intents to transform from data encoding to PCS. The ICC-absolute intent is implied by the media-relative intent, and the scaling described in Section 24.5 (PCS) of the ICC specification.

In the simplest BToAx tag structure the matrix and curve elements of the tag are not used (or are set to identity) and the CLUT converts directly from the PCS to the device encoding. In this case the input table is a uniform sampling of the entire PCS encoding, and the output values are computed using the inverse device model. More detail on LUT processing elements is given in Chapters 25 and 28.

The effective gamut of the device encoding is the result of transforming the PCS to the device encoding using the BToA1Tag and then converting these values back to the PCS using the AtoB1Tag. All PCS values that lie outside this effective gamut must be mapped to in-gamut PCS values, using a suitable gamut mapping algorithm. The gamutTag has the structure of a lutBToAType (in a v4 profile) or a lut8Type or lut16Type (in a v2 profile). For an input table which is a uniform sampling of the PCS encoding, the gamutTag stores an output table consisting of a zero where the PCS value is inside the effective gamut and a non-zero value for all PCS values outside the effective gamut.

The BToA1Tag uses the ICC colorimetric PCS and should be measurement based, so that all the in-gamut PCS colors have an output value corresponding to the device value with the smallest colorimetric error when produced on the device (in media-relative colorimetry). Similarly, the AToB1Tag should also be measurement based, so that all values in the device encoding are transformed by this intent to the corresponding PCS value with the smallest colorimetric difference (in media-relative colorimetry) from the actual measurement of the encoded value when produced on the device.

The other tags use the ICC perceptual PCS, and thus convert between a reference medium gamut with a physically realizable white and black point and (in v4 profiles) a well-defined gamut which ideally corresponds to the Perceptual Reference Medium Gamut described in the ICC.1:2010 specification. The PCS values which correspond to the data encoding should be compressed or expanded to this reference gamut, so that when v4 input and output profiles are combined, the two profiles use a common gamut. Further discussion of rendering intents can be found in Chapters 12, 13 and 25.

For each rendering intent, an AToBxTag should provide the "conceptual inverse" of the corresponding BToAxTag. More information on the inversion of profiles is given in Chapter 32.

24.13 Writing and Checking the Profile

When the profile header, tag table, and tag elements are completed and all elements correctly encoded, they are written to the binary format as described in the profile specification. The specified format for all tag types, including length, offsets, and the encoding of all elements within the tag type, must be observed precisely or the resulting profile will not be successfully parsed. Conformance of the profile with the ICC specification can be checked as described in Chapter 34.

Basics of an ICC Profile

ICC profiles contain required metadata, color processing data, and possible optional tags.

The metadata required in all profiles is a 128-byte header, a tag table listing the tags present in the profile and their locations, and the copyright, description, and media white point fields.

The basic steps in generating the color processing tags in a device profile can be summarized as Characterize–Adapt–Scale–Encode.

Characterize. Samples are generated and imaged on the device to be profiled. For an output device the samples will span the device data encoding, normally including the device primaries, while for an input device they will be the result of capturing a target which includes samples spanning the range of colors to be captured. Measurements of the samples are obtained and a model of the relationship between device encoding and colorimetry is generated. The appropriate combination of ICC color processing elements – curves, matrices, and CLUTs – to encode this relationship is selected.

Adapt. If the measurements are not relative to the D50 illuminant, they are chromatically adapted using a suitable chromatic adaptation transform. The 3×3 matrix which converts the media white under the original illuminant to the media white adapted to illuminant D50 is stored in the *chad* matrix.

Scale. All measurements (chromatically adapted to D50 if necessary) are normalized so that $Y_a = 1$ for a perfect reflecting diffuser, and scaled so that they are relative to the media white as follows:

$$X_{PCS} = \left[\frac{X_i}{X_{mw}}\right] X_a$$

$$Y_{PCS} = \left[\frac{Y_i}{Y_{mw}}\right] Y_a$$

$$Z_{PCS} = \left[\frac{Z_i}{Z_{mw}}\right] Z_a$$

where X_a is the measured X tristimulus value (after chromatic adaptation if required), X_{mw} is the X tristimulus value of the media white (also after chromatic adaptation if the original measurements were not relative to the D50 illuminant), and X_i is the X tristimulus value of the PCS white point as specified in the profile header. The values $[X_{mw}, Y_{mw}, Z_{mw}]$ correspond to those in the mediaWhitePointTag. Y_{PCS} and Z_{PCS} are computed similarly. Thus the PCSXYZ value of the media white is [0.9642, 1.0, 0.8249].

PCSLAB values are calculated from PCSXYZ using the 1976 CIELAB equations, except that X/X_n is replaced by X_{PCSXYZ}/X_i and similarly for Y and Z. Thus the PCSLAB value of the media white is [100, 0, 0].

Encode. The PCSXYZ or PCSLAB values are encoded in the appropriate tags, using the numeric type specified for the tag.

Creating a Display Profile

Version 4 display profiles can be either matrix based or LUT based. A matrix-based profile only includes a single rendering intent, which will normally be relative colorimetric. The PCS data will be encoded as PCSXYZ rather than PCSLAB.

A LUT-based display profile requires both AToB0 and BToA0 tags, and other rendering intents can optionally be provided. It is recommended that both colorimetric and perceptual intents are included in the profile, to provide color re-rendering to the display medium. The v4 profile format allows the PCS data to be either PCSXYZ or PCSLAB.

There are particular points to note in creating a display profile:

- Because the display peak white $(R = G = B = 255)$ is the media white, the PCSXYZ value of the media white matches the D50 illuminant. This value should also be encoded in the mediaWhitePointTag.
- The colorimetry of the display encoded in the profile should correspond closely to the stimulus actually observed by the user. This implies that measurements should include normal glare present in the viewing environment, but exclude measurement flare. A remote measuring instrument (e.g., a telespectroradiometer) positioned at the location of the user can be used to make such a measurement, but if a contact instrument is used instead an offset should be added to the data to allow for an estimate of the viewer-observed flare.

Tags for an example matrix-based display profile are shown below.

Tag	Size (bytes)	Value
"desc"	86	Matrix TRC v4 test profile[a]
"rXYZ"	20	[0.485 06, 0.250 11, 0.022 74][b]
"gXYZ"	20	[0.348 91, 0.697 80, 0.116 30]
"bXYZ"	20	[0.130 22, 0.052 09, 0.685 87]
"rTRC	2060	1024 values[c]
"gTRC"	2060	1024 values
"bTRC"	2060	1024 values
"wtpt"	20	[0.9642, 1.0, 0.8249][d]
"cprt"	78	"Colour Imaging Group, London"
"chad"	44	3×3 matrix[e]

[a]Encoded as Unicode string.
[b]Red primary after chromatic adaptation and scaling. Note that column sums of the three primaries are a close match to the white point.
[c]Encoded as curveType.
[d]For a display profile must match the PCS white point.
[e]Defines the conversion of the white point from original colorimetry to D50.

If the linearly additive model implied by the matrix-based profile is not suited to the actual device behavior, a LUT-based profile should be generated instead.

Creating a Printer Profile

In a v4 printer profile, lutBToAType and lutAToBType tags are provided for each of the three rendering intents. The additional matrix and curve elements add further functionality to this tag type, in comparison to the older lut8 and lut16 types.

Generating a LUT-based profile is discussed in detail in Chapter 28. In outline, the steps in generating a very basic A2B1 transform encoded in a lutAToBType are as follows:

1. Print and measure a test target which provides a sampling of the device data encoding. Suitable targets include those described in ANSI IT8.7/3 and IT8.7/4 and ECI (2002). The resulting measurement data will normally be D50, in which case chromatic adaptation is not required.
2. Normalize all data to the media white so that $L^* = 100$, $a^* = b^* = 0$ for the media white point.
3. Determine the curves required to linearize the device data with respect to the PCS. These can be encoded as the lutAToBType A curves in order to minimize errors in the CLUT.
4. Apply smoothing to the measurement data if appropriate.
5. Determine the characterization model to be used to calculate PCS values from the device encoding after linearization and smoothing.
6. Select the number of nodes to be used in the CLUT and generate an input table sampling the device data encoding.
7. Determine output values for each input entry using the device model and encode these as the lutAToBType CLUT.
8. Encode a linear curve in the lutAToBType B curves and leave the matrix and M curves empty.

To generate the corresponding B2A1 transform for the lutBToAType tag:

1. Determine any curves appropriate to the PCS side of the transform (e.g., to weight the PCS color space in favor of neutrals) and encode these as the lutBToAType B curves.
2. Determine the curves required to linearize the CLUT output to the device values (possibly inverting the lutAToBType A curves) and encode these as the lutBToAType A curves.
3. Determine the device model to predict output values from the B curves' output (possibly inverting the model used in generating the lutAToBType tag).
4. Select the number of nodes to be used in the CLUT and generate an input table sampling the PCS encoding.
5. For the CLUT nodes which are within the color gamut represented by the device encoding, compute output values using the device model.
6. For the CLUT nodes which are not inside the device encoding color gamut, select a suitable gamut mapping algorithm (such as HPMINDE) and apply this to compute in-gamut PCS colors; then using the device model compute output values.
7. The CLUT nodes which are not inside the device encoding color gamut are identified by using non-zero values in the CLUT encoded in the gamutTag.
8. Encode the CLUT output values as the lutBToAType CLUT.
9. Leave the matrix and M curves empty.

To generate the B2A0 and A2B0 tags, it is necessary to determine a suitable rendering from a perceptual reference medium to the device output. The ICC recommends using the Perceptual Reference Medium Gamut for this purpose. PCS encoding values which are outside the reference medium gamut will be clipped to this gamut, and then a rendering transform from the selected PRM to the gamut of the device encoding is determined. The goal of this rendering is to produce a pleasing reproduction on the output medium, and the PCS colorimetry may be adjusted as desired to achieve this.

The rendering can be determined for one direction of the transform, and the inverse then computed as the numeric inverse of that transform.

The tags in an example v4 output profile are shown below. The tags include a charTargetTag (signature "targ") which encodes the characterization data used in generating the device model used in creating the profile.

Tag	Size (bytes)	Value
"desc"	58	SWOP v4
"cprt"	92	FUJIFILM Electronic Imaging Ltd
"wtpt"	20	[0.7101, 0.7381, 0.5730]
"targ"	43 142	ANSI/CGATS TR001 measurement data file
"B2A0"	292 836	v4 lutBToAType with A curves, 3D CLUT, M curves, 3×4 identity matrix, and B curves
"A2B0"	28 296	v4 lutAToBType with A curves, 3D CLUT, and B curves
"B2A1"	292 836	v4 lutBToAType with A curves, 3D CLUT, M curves, 3×4 identity matrix, and B curves
"A2B1"	504 848	v4 lutAToBType with A curves, 3D CLUT, and B curves
"B2A2"	292 836	v4 lutBToAType with A curves, 3D CLUT, M curves, 3×4 identity matrix, and B curves
"A2B2"	28 296	v4 lutAToBType with A curves, 3D CLUT, and B curves
"gamt"	37 009	lut8Type with 3×3 identity matrix, input curves, CLUT, and output curves

25

ICC Profile Internal Mechanics

This chapter focuses on the engineering aspect of creating and interpreting ICC profiles. It covers some of the applicable mathematics (largely linear algebra) and some details on how ICC profiles may be encoded internally, such as differences between 8- and 16-bit encodings, and how to maintain precision.

It is well known among computer scientists that when real numbers are represented in a finite number of bits the result is usually an approximation. Consider this small piece of C code:

```c
int main(void)
{
  int i;
  float a = 0;
  for (i=0; i < 10; i++)
    a = a + 0.1;
  if (a == 1.0)
    printf("OK");
  else
    printf("Oops!");
  return 0;
}
```

On inspection, it appears obvious that the code should result in $a = 1$ and therefore the logical test should return true. However, when using floating point numbers, the result comes out at an approximation to 1 (differing by 1.11E-16 when computed at double precision). While the result is not significantly different for most practical purposes, the logical test fails.

It is important to understand such issues or your code may give unexpected and possibly incorrect results.

We will now consider some basic techniques which are a foundation for the issues covered in this chapter.

Color Management: Understanding and Using ICC Profiles   Edited by Phil Green
© 2010 John Wiley & Sons, Ltd

25.1 Rounding

In general, rounding is the process of reducing the number of significant digits in a number. The result of rounding is a "shorter" number having fewer non-zero digits, yet similar in magnitude. The result is less precise but easier to use.

ICC profiles do have a dependency on rounding, since they are binary digital files and as such are subject to quantization. Below we consider some rounding techniques.

25.1.1 Round-Toward-Nearest

As its name suggests, this algorithm rounds toward the nearest significant value. In many ways, this is the most intuitive of the various rounding algorithms, because values such as 5.1, 5.2, 5.3, and 5.4 will round down to 5, while values of 5.6, 5.7, 5.8, and 5.9 will round up to 6.

But what should happen in the case of a "half-way" value such as 5.5? The two options are to round it up to 6 or down to 5, and these schemes are known as round-half-up and round-half-down, respectively.

25.1.2 Round-Half-Up (Arithmetic Rounding)

This algorithm, which may also be referred to as arithmetic rounding, is the one that is typically associated with the concept of rounding. In this case, a "half-way" value such as 5.5 will round up to 6.

The problem with the round-half-up algorithm arises when we consider negative numbers. If positive values like +5.5 and +6.5 round up to +6 and +7, respectively, one would intuitively expect their negative equivalents of −5.5 and −6.5 to round to −6 and −7, respectively. In this case, we would say that our algorithm was *symmetric* (with respect to zero) for positive and negative values.

The direction "up" can be taken as referring to positive infinity, and based on this −5.5 and −6.5 would actually round to −5 and −6, respectively. We would class this as being an *asymmetric* (with respect to zero) implementation of the round-half-up algorithm:

Different applications perform rounding differently. For example, the round method of the Java Math Library provides an asymmetric implementation of the round-half-up algorithm, while the round function in MATLAB provides a symmetric implementation, and the round function in Visual Basic for Applications 6.0 actually implements the round-half-even algorithm.

25.1.3 Round-Half-Even

If half-way values are always rounded in the same direction (e.g., if +5.5 rounds up to +6 and +6.5 rounds up to +7, as is the case with the round-half-up algorithm described above), the result can be a bias that grows as more and more rounding operations are performed. One solution toward minimizing this bias is to sometimes round up and sometimes round down.

In the case of the round-half-even algorithm (which is often referred to as "Bankers' Rounding" because it is commonly used in financial calculations), "half-way" values are rounded toward the nearest even number. Thus, $+5.5$ will round up to $+6$ and $+6.5$ will round down to $+6$.

This algorithm is, by definition, symmetric for positive and negative values, so both -5.5 and -6.5 will round to the nearest even value, which is -6.

In the case of data sets that feature a relatively large number of "half-way" values, the round-half-even algorithm performs significantly better than the round-half-up scheme in terms of total bias. It is for this reason that the use of round-half-even is a legal requirement for many financial calculations around the world.

25.2 Converting Between Domains

Let us assume we have two different domains, say $[0,1]$ and $[0,255]$, both beginning with zero. If we want to convert a value from source domain to destination domain, then in principle this is given by

$$Y = \frac{\text{destination domain max}}{\text{source domain max}} \times X. \qquad (25.1)$$

For example, 0.5 in domain 0.1 is converted to

$$\frac{255}{1} \times 0.5 = 127.5. \qquad (25.2)$$

Now consider converting from 8 bits to 16 bits. In this case we go from $[0,255]$ to $[0,65535]$, which following the example of Equation (25.1) gives

$$\frac{65\,535}{255} \times X = Y. \qquad (25.3)$$

However, this returns 257, rather than 256. So to convert an 8-bit value to a 16-bit one, we see that is necessary to multiply by 257. We can also get the same result and improve the computational performance by the following:

$$X \times 257 = X \times 256 + X = (X \ll 8)|X. \qquad (25.4)$$

This is faster to compute, as it only uses a shift left plus bitwise OR operators.

But what happens in the inverse situation, where we need to convert from 16 bits to 8 bits? The result of Equation (25.3) suggests that we need to divide by 257. However, division is computationally expensive compared to addition and bitwise operations. Using a simple shift right $X \gg 8$ would be wrong, as it would be equivalent to dividing by 256. A faster method of doing the division by 257, which also includes arithmetic rounding, is

$$Y(((X \times \text{0xFF01}) + \text{0x800000}) \gg 24) \,\&\, \text{0xFF}. \qquad (25.5)$$

Slightly more complex is the situation where both domains do not begin with zero. The expression can be computed as a line that goes across two points and the result is still very simple:

$$y = \alpha \times x + \beta$$

$$\alpha = \frac{Dest\max - Dest\min}{Src\max - Src\min} \qquad (25.6)$$

$$\beta = Dest\min - \alpha \times Src\min$$

where *Dest* and *Src* refer to the destination encoding and transform source respectively. For example, from $-127\ldots+128$ to $0\ldots65\,535$,

$$\alpha = \frac{65\,535}{128 - (-127)} = 257$$

$$b = 0 - 257 \times (-127) = 32\,639 \qquad (25.7)$$

$$y = 257 \times x - 32\,896$$

Now we can check the end points to confirm that the results are as expected:

$$Y(-127) = 257 \times (-127) + 32\,639 = 0$$

$$Y(+128) = 257 \times 128 + 32\,639 = 65\,535. \qquad (25.8)$$

25.3 Fixed Point

The idea behind fixed point mathematics is that we assume that a decimal point is present even though it is not explicitly included.

Let's take for example this approximation of π: 3.141 59.

A fixed point number is an integer that represents a number consisting of a whole part (e.g., 3) and a fractional part (e.g., 0.141 59). The format of a fixed point number is usually described as "$m \cdot n$," where m is the number of bits in the whole part and n is the number of bits in the fractional part; $m + n$ is the total number of bits. For example, "8.24" means a 32-bit integer with an 8-bit whole part and a 24-bit fractional part. Fixed point formats may have signs too.

To convert a number to fixed point it is multiplied by 2^n. To convert the fixed point number back, it is divided by 2^n. Hence to represent 3.141 59 as an 8.24 fixed point number, we multiply it by 2^{24} resulting in 52 707 134. To return to the decimal representation, we divide by 2^{24} which gives 3.141 589 999 as the result.

25.3.1 Fixed Point Mathematics

Fixed point mathematics is somehow surprising. To understand why it works, consider that if you do all of you mathematics without the decimal point, and then add the decimal point, you get the same result.

For example,

$$1.2 \times 3.4 = 4.08. \tag{25.9}$$

If the decimal point is removed:

$$12 \times 34 = 408. \tag{25.10}$$

That implies some interesting properties. Suppose A, B, and C are the fixed point versions of a, b, and c:

$$A = a \times 2^n$$
$$B = b \times 2^n. \tag{25.11}$$

Addition: if $c = a + b$, then

$$C = (a+b) \times 2^n = (a \times 2^n) + (b \times 2^n)$$
$$C = A + B. \tag{25.12}$$

In other words, to perform the addition of two fixed point numbers, you just add their fixed point representation. *Subtraction* works the same way.

Multiplication: if $c = a \times b$, then

$$C = (a \times b) \times 2^n = (a \times 2^n) \times (b \times 2^n)/2^n$$
$$C = A \times B/2^n. \tag{25.13}$$

So to multiply two fixed point numbers, you multiply them and then divide by 2^n (or shift right n bits). Note that overflow is a problem that has to be dealt with.

Division: if $c = a/b$, then

$$C = (a/b) \times 2^n = (a \times 2^n)/(b \times 2^n) \times 2^n$$
$$C = A/B \times 2^n. \tag{25.14}$$

In other words, to divide two fixed point numbers, you divide them and then multiply by 2^n (or shift left n bits). Note that underflow is a very serious problem which is usually dealt with by rearranging the order of operations like this:

$$C = A \times 2^n / B. \tag{25.15}$$

Again, you have an overflow problem that has to be dealt with. The easiest and safest way to deal with overflow is to use a larger integer size to store intermediate values. For example, you might use 64-bit numbers to do 8.24 fixed point mathematics.

The drawback is that it can be slower, or there might not be a larger integer size available. You could also find a way of rearranging the calculation (as in the example of the divide operation in Equation (25.15)), but that could have drawbacks.

One issue to consider with fixed point encodings is that a number in decimal notation may be approximated when represented in fixed point binary. Consider, for example, the encoding of gamma 2.2 in a RedTRC tag. This particular tag uses an 8.8 fixed point type. The encoded number may be computed by multiplying by 2^8 (256), which gives $2.2 \times 256 = 563.2$, which after rounding gives 563. Now when we try to recover the original value

$$563/256 = 2.1992,$$

it can be seen that 2.2 cannot be encoded exactly in 8.8 fixed point format. While this is the case when using the gamma encoding, in v4 of the ICC specification there are parametric curves with a 15.16 encoding for parameters. Using this encoding, 2.2 would turn into $2.2 \times 65\,536 = 144\,179.2$, still not exact but the roundtrip is $2.199\,997$, which is much closer to 2.2. Parametric curves should therefore be used wherever possible.

25.4 Interpolation

Interpolation can be thought of as an "educated guess," where a value is calculated based on the values of its neighbors. Linear interpolation is computationally fast and easy to implement, but it is not very precise when the underlying function is not linear at the point being interpolated.

The simplest form of interpolation is across a single segment, defined by the two values which lie at the ends of the segment. Again this is just a linear algebra problem:

$$Y = \alpha x + \beta.$$

We have two known points (X_a, Y_a) and (X_b, Y_b). So, solving the system

$$Y_a = \alpha X_a + \beta$$

$$Y_b = \alpha X_b + \beta$$

gives

$$y = y_a + \frac{(x - x_a)(y_b - y_a)}{(x_b - x_a)}.$$

Where the underlying function is a straight line, this produces an exact value. For a nonlinear function, we break the curve into a set of linear segments, such that for each segment we have two known end points. We can interpolate the value of any 1D function using this technique, and the greater the number of nodes, the more precise the approximation.

A curve can thus be represented by a table of the nodes for which we have known output values. To interpolate using such a table, for any given input we have first to select (or "extract") the surrounding nodes. For a table encoded at 16-bit precision we have a value that goes from 0 to 65 535 and want a value that goes from 0 to the maximum number of nodes. This is just a matter of changing domain, as described in Section 25.3 above.

It is sometimes stated that a particular number of nodes has inherent advantages. Consider a table indexed in 16 bits where we want to convert to nodes $n - 1$, since there is one final node at the very end:

$$y = \frac{n - 1}{65\,536} x = \frac{n - 1}{3 \times 5 \times 17 \times 257} x.$$

While it may appear efficient to use the most significant bits of the index to address nodes, this approach is wrong. This can be seen by inspecting the last indexing value: 0xFFFF in our sample should map to the last node with 0 as offset, but, instead, the result is the anterior node and an offset of 255. We can still accommodate this by converting domains, from 0...0xFFFF to 0...0x10000 in our example – that is, applying a factor of 65 536/65 535 to the index before extracting nodes. This changes the equation to

$$y = \frac{n - 1}{65\,536} x' = \frac{n - 1}{2^{16}} x'.$$

Consider what happens when we use, say, 17 nodes:

$$y = \frac{17 - 1}{65\,536} x' = \frac{16}{2^{16}} x' = \frac{2^4}{2^{16}} x' = \frac{x'}{2^{12}}.$$

Dividing by 2^{12} can be accomplished by shifting right $\gg 12$ bits, so this is the same as taking the $16 - 12 = 4$ most significant bits (MSBs) of the index.

So for any $n - 1$ equal to powers of two you can get the nodes by simply taking the most significant bits:

- For 17 nodes use 4 MSBs
- For 33 nodes, use 5 MSBs
- For 257 nodes, use 8 MSBs.

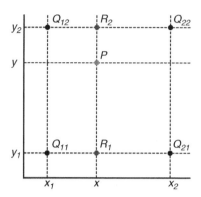

Figure 25.1 Bilinear interpolation

It is still necessary to apply the 65 536/65 535 factor described above, and in practice an extra bit is needed to store the temporary result (17 bits for a 16-bit index in our example).

25.4.1 Bilinear Interpolation

The one-dimensional interpolation discussed above can readily be extended to two dimensions, where the number of nodes that affect the point of interest is now four rather than two (Figure 25.1).

The key idea is to perform linear interpolation first in one direction and then in the other direction. Suppose that we want to find the value of the unknown function f at the point $P = (x, y)$, where we know the value of f at the four points $Q_{11} = (x_1, y_1)$, $Q_{12} = (x_1, y_2)$, $Q_{21} = (x_2, y_1)$, and $Q_{22} = (x_2, y_2)$.

If we choose a coordinate system in which the four points where f is known are $(0, 0)$, $(0, 1)$, $(1, 0)$, and $(1, 1)$, then the interpolation formula simplifies to

$$f(x,y) \approx \begin{bmatrix} 1-x & x \end{bmatrix} \begin{bmatrix} f(0,0) & f(0,1) \\ f(1,0) & f(1,1) \end{bmatrix} \begin{bmatrix} 1-y \\ y \end{bmatrix}.$$

While bilinear interpolation is not used on ICC profiles, the logical extension to three or more dimensions is certainly used.

25.4.1.1 Trilinear Interpolation

Trilinear interpolation (Figure 25.2) is perhaps the simplest method, and is based on the extension of the one- and two-dimensional methods. Again we do interpolation in one direction and then the remaining ones in turn.

The computational cost of interpolation increases with the number of nodes, and different techniques are used to optimize the process by minimizing the number of nodes that are used

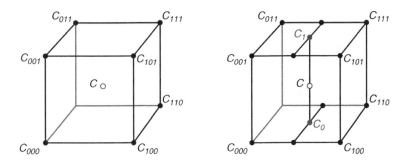

Figure 25.2   Trilinear interpolation

in the computation. One commonly used method is tetrahedral interpolation. In both tetrahedral and trilinear interpolation the grid points along each of the three axes serve to divide the volume into a set of rectangular hexahedra (Figure 25.3). To interpolate at a particular point, the first step is determine in which hexahedron the point lies. In tetrahedral interpolation, the hexahedron is further subdivided into six tetrahedra. There are many possible subdivisions into tetrahedra, but there is one such subdivision that is typically used for color space conversion.

Once the tetrahedron containing the interpolation point is identified, the interpolation is computed as a weighted sum of the grid values at the vertices of that tetrahedron.

If one examines each of the tetrahedra, one can see that they all share a common edge on the diagonal from $(0, 0, 0)$ to $(1, 1, 1)$. This is the reason that this particular tetrahedral subdivision is typically used for color space conversions. In an RGB space, this diagonal contains the neutral

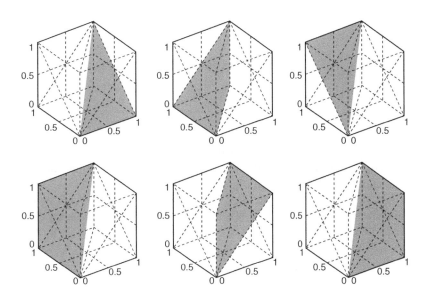

Figure 25.3   Tetrahedral subdivisions of the cube

(or gray) colors, which are particularly important in color reproduction. Tetrahedral interpolation can thus be used to convert colors in a more accurate and pleasing way, as well as
having a lower computational cost.

25.5 Encoding of Some Common Color Spaces

With the basic tools introduced above, we can now explore how color spaces are encoded.

25.5.1 RGB

Scientific applications often employ a [0,1] range for RGB. A [0,255] range is perhaps the most
widely used for historic reasons, although decimal numbers are also sometimes found. Here we
are going to use a 16-bit profile, so that we can apply the rules described above for converting
between domains:

$$0 \ldots 1.0 \rightarrow 0 \ldots 0xFFFF : rgb \times 65\,535$$

$$0 \ldots 255.0 \rightarrow 0 \ldots 0xFFFF : rgb \times 257.$$

25.5.2 CMYK

CMYK is slightly different than RGB since it is most commonly used to represent a percentage
of colorant, such that 100% is the maximum amount of colorant. So, applying the same rules as
for RGB:

$$0 \ldots 1.0 \rightarrow 0 \ldots 0xFFFF : CMYK \times 65\,535$$

$$0 \ldots 100.0 \rightarrow 0 \ldots 0xFFFF : CMYK \times \frac{65\,535}{100} = CMYK \times 655.35$$

25.5.3 XYZ

For XYZ, the ICC uses a notation normalized to 1.0, so D50 is represented as (0.9642, 1.0,
0.8249) instead of the more usual (96, 100, 82). To fit them in 16 bits, the values are encoded in a
fixed point notation, specifically 1.15 unsigned.

What is then the encodeable range? Since real zero equals zero in fixed point, and the
maximum encoded number would be 0xFFFF in 1.15 fixed point, we can convert this back to a
real number by dividing by 2^n

$$65\,535/2^{15} = 1.999\,969\,482\,421\,875.$$

This is almost, but not quite, two.

25.5.4 CIELAB

Things get more complicated when dealing with CIELAB. XYZ allows only 16-bit encoding, while CIELAB can be encoded in either 8 or 16 bits. Unfortunately the ICC specification has two different, incompatible ways to encode CIELAB in 16 bits. We will consider each one in detail.

25.5.5 Lab8

The first encoding was for v2 8-bit profiles. Since in CIELAB the a^*, b^* part can be negative, we need a means to encode this. It would be possible to use one bit for the sign or just change domain as shown in Table 25.1. Using a bit for the sign has the unwelcome side effect that since signed numbers are using two's complement, values close to zero in both the positive and negative sides are quite different when encoded and the transition across zero is not smooth.

This leads to very poor results in interpolation. Instead, we can use a domain translation, so we define our destination domain as [0,255] (remembering that this is an 8-bit encoding), the source domain as [0,100] for L^*, and 127...−128 for a^*, b^*. This gives rise to the following conversions.

For L^*:

$$L^*_{\text{encoded}} = 255/100 \times L^*_{\text{unencoded}}$$

$$L^*_{\text{unencoded}} = 100/255 \times L^*_{\text{encoded}}$$

and then for the a^*, b^* part:

$$ab_{\text{encoded}} = ab_{\text{unencoded}} + 128$$

$$ab_{\text{unencoded}} = ab_{\text{encoded}} - 128.$$

There is at least one major flaw in this design: zero is not centered since the a^*, b^* range goes from −128 to 127. This is important because CIELAB is almost always used as the PCS in a

Table 25.1 Domain change for signed 8-bit values

a^*/b^* to encode	Signed value encoded as:	Change domain:
−2	0xFE	0x7E
−1	0xFF	0x7F
0	0x00	0x80
1	0x01	0x81
2	0x02	0x82

three-dimensional LUT profile, and so it will be used as the index for our LUTs. Suppose we want to place all values on the neutral axis onto nodes in the LUT; then $a^* = b^* = 0$ in the encoded notation is

$$ab_{encoded} = ab_{unencoded} + 128 = 0 + 128 = 128.$$

What node configuration would ensure that our neutral axis $a^* = b^* = 0$ falls on specific nodes?

$$\frac{n-1}{255} \times 128 = \frac{n-1}{3 \times 5 \times 17} \times 2^7.$$

All the numbers in the divisor are prime factors, so the only possible configuration for n is $3 \times 5 \times 17 + 1 = 256$. In practice this is far too great a number of nodes for a three-dimensional grid.

25.5.6 Lab16 (v2)

For 16 bits, a particular encoding was described in the v2 ICC specification. The format of this encoding was defined in the same way as Lab8, but in fixed point, the upper byte is the same as in Lab8 and the lower byte contains the added fractional part. As seen in the fixed point discussion above, we can convert to fixed point 8.8 encoding by multiplying by 256.

Let us consider how some common numbers are encoded in this particular format:

$$L \times 100 \rightarrow 0xFF00 = 65\,280$$

$$a/b = 0 \rightarrow 0x8000 = 32\,768.$$

Unfortunately, this approach makes the centering problem worse. In order to have $L^* = 100$ (the PCS white point) on an exact node:

$$\frac{n-1}{65\,535} \times 65\,280 = \frac{n-1}{3 \times 5 \times 17 \times 257} \times 3 \times 5 \times 17 \times 2^8 = \frac{256}{257} \times (n-1).$$

Thus only 256 nodes would meet the requirement of placing $L^* = 100$ on a node, and for the resulting a^*, b^* part

$$\frac{n-1}{65\,535} \times 32\,768 = \frac{n-1}{3 \times 5 \times 17 \times 257} \times 2^{15}.$$

There is simply no way to center this.

25.5.7 Lab16 (v4)

Version 4 of the ICC specification tried to fix this by redefining the Lab16 encoding as Lab8 times 257. This means that we can convert between 8 and 16 bits in the same way for other color spaces. As a result, $L^* = 100$ is now encoded as 0xFFFF, which is always the last node, and this applies regardless of the number of nodes.

For the a^*, b^* part we still have an issue with neutrals. Zero now is encoded as 0x8080 32 896 and

$$\frac{n-1}{65\,535} \times 65\,280 = \frac{n-1}{3 \times 5 \times 17 \times 257} \times 257 \times 2^7 = \frac{128}{3 \times 5 \times 17} \times (n-1).$$

It is thus still not possible to make $a^* = b^* = 0$ fall exactly on a node. Workarounds for this problem exist but are beyond the scope of this chapter.

25.6 LUT Types

The current ICC specification defines four different LUT types, plus the multi-processing elements (MPE) LUTs. In the four non-MPE LUT types we have LUT8, LUT16, LutAToB, and LutBToA. The LUT8 and LUT16 types were the only LUTs defined in versions previous to v4, while LutAToB and LutBToA are in v4 and above.

Where precision is important, LUT8 in the BToAx direction should be avoided whenever possible. This is because the allowed PCS only in this case is Lab8, and rounding leads to a digital count of a difference of approximately $1\Delta E$. That is not the case for Lab16, which has only about $0.004\Delta E$ rounding error. An LUT16 table is twice the size of an LUT8 table, but has 250 times more precision.

LUT8 and LUT16 have the following processing elements: matrix – one-dimensional curves – CLUT – one-dimensional curves.

In LUT8 and LUT16, the matrix element can only be used when PCS is XYZ. In LUT8, both curves are fixed to 256 entries.

LutAToB have the elements: A curve – CLUT – M curve – matrix – B curve. However, only certain combinations are allowed:

B
M – matrix – B
A – CLUT – B
A – CLUT – M – matrix – B.

LutBToA has the same elements but in the reverse order:

B curves – matrix – M curves – CLUT – A curves.

In this case, the allowed combinations are:

B
B – matrix – M
B – CLUT – A
B – matrix – M – CLUT – A.

25.6.1 Matrices

One of the many possible uses of matrices is to use a form of RGB as an intermediate space. Since the PCS can be XYZ, we can use a matrix to convert from PCSXYZ to a gamma 1.0 RGB. This has the advantage of no negative numbers and a very large gamut. The resulting RGB may then be accommodated to a more perceptually uniform RGB by means of the one-dimensional curves. Finally we can use the CLUT to fine-tune the RGB obtained from the matrix and curves. With this approach we can map white and black on a node, and let the CMM do the necessary clipping on out-of gamut XYZ values before even entering the pre-linearization.

25.6.2 Pre-linearization Curves

Pre-linearization curves are applied to the data before it enters the CLUT. They are very useful in optimizing the *Lab* encoding to the three-dimensional CLUT. Using 258 entries, we can map the 0xFF00 white of Lab16 to the upper node. We can also center $ab = 0$ on the gray axis. Furthermore, we can implement different node densities by using nonlinear curves. Consider the curve configuration in Figure 25.4. The gray line is the pre-linearization curve applied to L^*, while the black curve is applied to a^*, b^*. This sigmoidal curve has the effect of increasing the resolution near the neutral axis at the expense of highly saturated colors. This is an effective strategy as highly saturated colors are likely to be outside the device gamut, and will suffer less from reduced resolution than would a neutral color.

Another use would be to decouple gamma. If we want to create a device link that goes from AppleRGB (with a gamma of 1.8) to sRGB, which has a gamma near 2.2, we can use a 1.2 exponential as pre-linearization. Then when the curve is applied to AppleRGB the gamma is adjusted and we can then complete the transform with a simple 3×3 matrix.

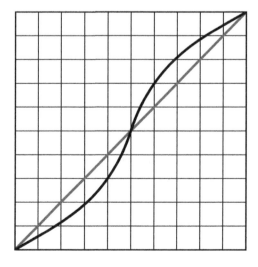

Figure 25.4   Pre-linearization curve

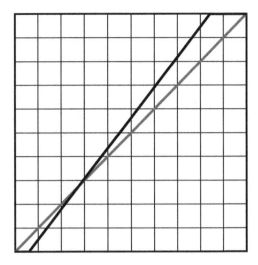

Figure 25.5   Post-linearization curve

25.6.3 Post-Linearization Curves

Post-linearization curves are applied to the output of the CLUT, and are available in both v2 and v4. Here we will consider only one of the multiple uses of these curves (Figure 25.5).

Suppose we are implementing relative colorimetric intent. For the sake of simplicity in this example we choose an RGB space with a black point in CIELAB $= (10, 0, 0)$. Even using pre-linearization curves, we can have a situation where the black point falls between two nodes (say node 0 and node 1). There is no problem in computing RGB for node 1, which would be RGB $= (5, 5, 5)$. However, node 0 is now below the real black point and the theoretical RGB value is out of gamut: $(-4, -4, -4)$. Since an LUT cannot contain negative numbers we clamp it to zero.

Now, what happens when the black point $Lab = (10, 0, 0)$ arrives at the LUT? Instead of the expected $(0, 0, 0)$ the result is $(2, 2, 2)$, because the interpolation segment goes from $(0, 0, 0)$ to $(5, 5, 5)$.

This is a significant issue in certain situations such as black point compensation, since the effect is amplified and any shadow detail is lost.

This problem can be solved by post-linearization, using the curve to add an offset for all values. In our example that is -4 for each RGB component. Then the values stored in the nodes are $(0, 0, 0)$ and $(9, 9, 9)$. When interpolation takes place, $L^* = 10$ is interpolated between 0 and 9, giving 4, then the curve subtracts 4, and the result is 0. For an L^* falling on node 1, the value is still correct, since $9 - 4 = 5$, and an absolute zero value is already clamped to zero by the curve.

25.6.4 Multi-processing Elements

A new structure, MPE, for multi-processing elements, is being added to the ICC specification. When this structure is used it greatly simplifies the task of profile creators while making things a little bit harder for CMM implementers. For certain applications it provides a significant improvement in functionality. An MPE type is a container that can hold any combination of curves, matrices, and LUTs, which can be encoded using floating point. So, no fixed point tricks, no limited precision? Not really. This implies that we are no longer restricted to a 0...0xFFFF range but can encode values from negative infinity to positive infinity, and only at the very last stage does the CMM clamps the values. At present the MPE structure is limited to DToB/BToD tags, but it may evolve to incorporate new types in the future. Curves in MPE may be specified in several segments, and each segment may be sampled or specified as a formula. This is a very powerful approach, but it has some complexities that need to be fully understood if it is to be used successfully.

26

Use of the parametricCurveType

In Version 4 of the ICC profile specification, parametricCurveType was introduced as an alternative to curveType for the representation of one-dimensional transfer functions. Either type can be used for the TRC tags or for the A curves, B curves, and M curves embedded in lutAtoBType or lutBtoAType tags. In contrast to the older curveType, the new v4 type defines curves by closed-form expressions, rather than by one-dimensional LUTs. Each curve is a scalar function of a scalar variable, but the expressions also involve constants, or *parameters*, which are encoded in the corresponding profile tags.

The v4 profile specification supports five different *function types*, requiring between one and seven parameters. The specification places no restrictions on the values of these parameters, aside from those imposed by the format. According to Clause 10.15, the parameters are encoded in the s15Fixed16Number format. Thus, the values can range from $-32\,768$ to almost $+32\,768$ (actually $32\,768 - 1/65\,536$) in steps of $1/65\,536$, or $0.000\,015\,258\,789\,062\,5$. These restrictions are quite mild and, in practice, are hardly noticeable.

The parametricCurveType can be used to encode a wide variety of different functions. Profiles using the parametricCurveType can contain a wide range of possible parameter sets, and if care is not taken with their selection, some possible parameter sets can create computational problems for a CMM. The purpose of this chapter is to call attention to these problems, which can include divide-by-0 faults, complex roots, discontinuities in value and slope, and inversion ambiguities.

A CMM developer is faced with difficult decisions on how best to handle these problems. Different choices may be made by different developers, which can lead to inconsistent results among CMMs. Some of these choices may even interfere with legitimate choices made by profile creators.

The ICC specification provides guidance on avoiding complex or undefined values in one particular case, through the selection of the parameter d. Other issues can arise in the implementation of the parametricCurveType, and this chapter aims to provide some further guidance in this context for both the profile creator and the CMM developer.

Color Management: Understanding and Using ICC Profiles   Edited by Phil Green
© 2010 John Wiley & Sons, Ltd

26.1 Fundamentals

Curves defined by the parametricCurveType are scalar functions of a scalar variable, which can be written

$$y = f_n(x)$$

with $n = 0, 1, 2, 3,$ or 4 to designate the different function types. The domain of these functions is the unit interval $[0, 1]$ – that is, $0 \leq x \leq 1$. The range is also $[0, 1]$, so that y will be clipped if the closed-form expressions produce a value that is outside that interval.

In some cases, the CMM will need to evaluate the inverse of a parametric curve. We will write the inverse of $f_n(x)$ as follows:

$$x = F_n(y).$$

Here, too, the domain and range are $[0, 1]$, and clipping will occur if necessary. The inverse function will be needed when a TRC tag is evaluated in the PCS-to-device direction. (In the case of embedded curves, the profile creator provides the inverse in a separate tag.)

All five function types are variations on the basic power law

$$y = x^\gamma,$$

where γ ("gamma") is a constant parameter. They differ from one another in the use of various factors or offsets applied to x or y or in the partitioning of the domain into segments where other expressions are applied. The expressions are based on those typically used to model the transfer characteristics of CRT monitors or to define standard color spaces.

In mathematical terms, the parametric curves are normally used to define real-valued, continuous, smooth, monotonically non-decreasing functions mapping $[0, 1]$ onto $[0, 1]$. Figure 26.1 shows a typical example: the electro-optical transfer function used in HDTV, as

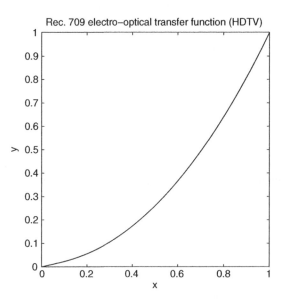

Figure 26.1   Well-behaved curve

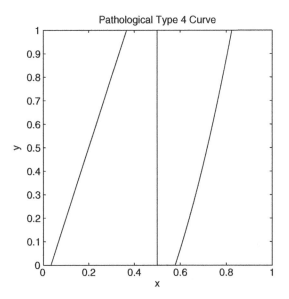

Figure 26.2   Badly behaved curve

defined by Recommendation ITU-R BT.709. This is a type 3 curve with a short linear segment at the origin, which becomes tangent to the power-law curve at $x = 0.081$.

However, any given parameter set may have a different effect and may, in some cases, result in a function exhibiting unwanted or even pathological behavior. See Figure 26.2 for an extreme example. This is a legitimate type 4 curve in which the parameters have been assigned values such that the linear segment and a power-law segment are both clipped and are discontinuous with each other, with a large reversal. This curve has no inverse.

While this curve has been specially constructed to illustrate bad behavior and is unlikely to occur in practice, a CMM has to be prepared for every eventuality.

26.2 The Power-Law Exponent

Typically, the exponent γ will be a small positive number. If so, x^{γ} will increase monotonically and will map 0 to 0 and 1 to 1. The inverse, $y^{1/\gamma}$, will have the same properties. See Figure 26.3 for examples.

If $\gamma < 0$, however, the function values for $0 < x < 1$ will decrease monotonically and will all be greater than one and, therefore, will all be clipped to one. The function diverges as x approaches zero, so the CMM must be careful of overflow conditions (or even divide-by-0 faults) occurring before the clipping stage. The inverse is ambiguous. (The section on inverse evaluation below has more discussion on this point.) See Figure 26.4, where $\gamma = -0.5$.

If $\gamma = 0$, the function is identically equal to one. The inverse is undefined, and even its exponent, $1/\gamma$, is undefined.

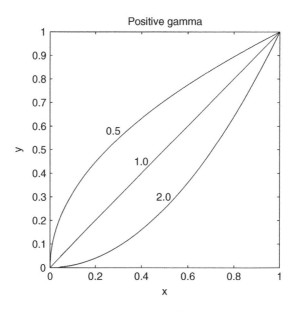

Figure 26.3   Power law with exponent $\gamma > 0$

It is not obvious how (or even whether) a CMM should process data when γ is zero or negative. The profile is almost certainly corrupt or in error in such a case. There is clearly no point in defining a constant or decreasing function which will be clipped over the entire domain. The CMM developer may choose, in such a case, to reject the profile or to replace

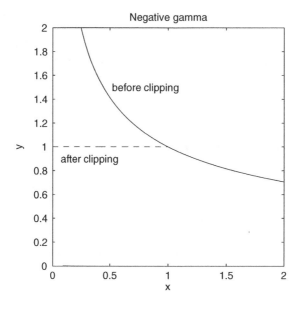

Figure 26.4   Power law with exponent $\gamma = -1/2$

the parametric curve with the identity function, $y = x$ (which amounts to setting $\gamma = 1$), or some other default. Ideally profile creators should abide by the condition $\gamma > 0$, and CMM developers should treat any occurrences of negative and zero values as errors and take appropriate action.

26.3 The Power-Law Argument

Function type 0 is just the basic power law described above:

$$y = f_0(x) = x^\gamma$$

$$x = F_0(y) = y^{1/\gamma}.$$

In the other types, the argument to the power law is not simply x, but a linear expression in x. In types 1, 2, 3, and 4, the argument, which we will call s, takes the form

$$s = ax + b,$$

where a and b are additional parameters. The power law is then s^γ. Since there is no practical restriction on the parameter values, s can take on any value as x varies between 0 and 1. If s is negative, s^γ can be imaginary or complex. (It will be real for integer γ, but γ cannot be restricted to integer values.) In such a situation, a CMM might choose to take the real part of the expression. Alternatively, it could take the absolute magnitude. It could arbitrarily set the expression to zero or one. Another option is simply to require s to be non-negative.

In types 1, 2, 3, and 4, the domain is divided into two segments, and the power law is employed only in the higher segment. For instance, the definition of type 1 is

$$y = f_1(x) = 0, \quad 0 \le x < -b/a$$
$$= s^\gamma, \quad -b/a \le x \le 1.$$

In normal usage, a will be positive and b will be negative, so that the segment boundary, $-b/a$, occurs at a positive value of x. The function is identically zero in the lower segment (Figure 26.5).

The argument s is non-negative throughout the higher segment, where the power law is in effect:

$$(-b/a \le x) \Rightarrow (0 \le ax + b = s).$$

This conclusion is verified by multiplying both sides of the first inequality by a and then adding b; it holds only if $a > 0$, however. Indeed, the inequality is reversed for negative a. (And if $a = 0$, the segment boundary itself is indeterminate.) It seems reasonable to impose

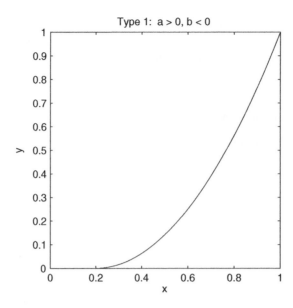

Figure 26.5 $f_1(x)$, with boundary at 0.2

the condition $a > 0$ as a requirement, so negative and zero values can then be treated as errors, and the CMM can take appropriate action – for instance, by substituting $a = 1$ or some other default. But, as in the case of γ, the CMM developer needs to be reassured that profile creators do not have a legitimate use for negative or zero values.

Another reason to require that a be positive is that it compels the power-law function to be monotonically non-decreasing, which is the normal case.

Similarly, one might consider imposing the condition $b < 0$. However, this is probably not necessary. Positive values of b simply mean that the segment boundary will occur at negative x. The power law will be in effect over the entire domain, and there will be no lower segment with $y = 0$. In such a case, s will always be positive, and there will be no risk of complex values. (See Figure 26.6.)

Note that, if the profile creator's intention is to have $y = 1$ just where $x = 1$, then the parameters should meet the condition $a + b = 1$.

Type 2 curves have a similar structure, with a segment boundary at $x = -b/a$, so the same analysis applies. However, in types 3 and 4, the segment boundary is defined by an independent parameter, d. In the absence of restrictions, it is quite possible for d to be less than $-b/a$. There can then be values of x in the higher segment $(x > d)$ at which $s = ax + b$ will be negative. The power law may then produce complex numbers. Figure 26.7 shows such an example. Here the absolute magnitude has been taken of the complex y values in the interval between d and $-b/a$, but that is an arbitrary choice. A CMM may, just as arbitrarily, take the real part of the y values or set them to zero. Since the occurrence of complex values is unlikely to be intentional, there is no universally correct way to handle them.

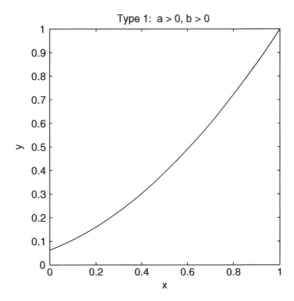

Figure 26.6 $f_1(x)$, with boundary at negative x

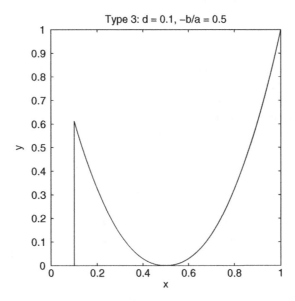

Figure 26.7 $f_3(x)$, with absolute magnitude of complex y values

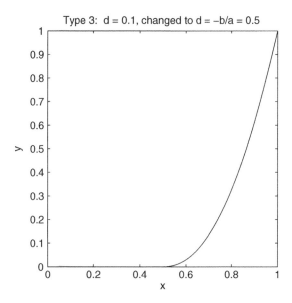

Figure 26.8 $f_3(x)$, with complex y values eliminated

To prevent complex values occurring, the CMM could reject such a value of d and replace it with $d = -b/a$. (See Figure 26.8.) The ICC recommends that $d \geq -b/a$ and profile creators should be aware that CMMs may impose this condition.

26.4 Continuity

Continuity across the segment boundary is guaranteed for types 1 and 2. For type 1 (see definition above), the value of s is exactly zero at the boundary, so the power law will yield a value of zero there. Below the boundary, in the lower segment, the function is identically zero, so $f_1(x)$ is continuous by definition.

Type 2 is similar to type 1, with the addition of an offset:

$$y = f_2(x) = c, \quad 0 = x < -b/a$$
$$= s^\gamma + c, \quad -b/a \leq x \leq 1,$$

where c is a constant parameter. The function is identically equal to c in the lower segment, and the power-law curve (in the upper segment) starts out at c, so $f_2(x)$ is also guaranteed to be continuous at the segment boundary. Figure 26.9 shows a typical example.

Types 3 and 4 do not enjoy a similar guarantee. Here is the definition of type 3:

$$y = f_3(x) = cx, \quad 0 \leq x < d$$
$$= s^\gamma, \quad d \leq x \leq 1.$$

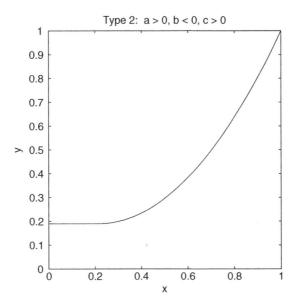

Figure 26.9 $f_2(x)$ with positive vertical offset

The function will be continuous at $x = d$ only if

$$cd = (ad + b)^{\gamma}.$$

Figure 26.10 shows a curve that violates this condition.

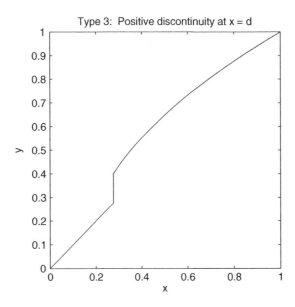

Figure 26.10 $f_3(x)$, discontinuous

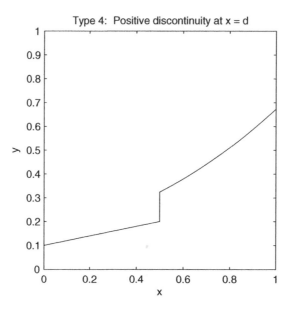

Figure 26.11 $f_4(x)$, discontinuous

Type 4 is defined as follows:

$$y = f_4(x) = cx + f, \quad 0 = x < d$$
$$= s^\gamma + e, \quad d \le x \le 1$$

where e and f are additional parameters. The corresponding continuity condition is

$$cd + f = (ad + b)^\gamma + e,$$

a relation involving seven parameters. Figure 26.11 shows a curve violating this condition.

Clearly, the profile creator should be aware of these conditions. In most, if not all, cases, the intention will be to encode a continuous function. Minor discontinuities may well occur through rounding of the parameter values, however, and they can be of either sign, so that reversals (non-monotonic behavior) will occur if the discontinuity is negative. Care is needed in the computation of the parameters if continuity problems are to be avoided.

Discontinuities in themselves do not cause computational problems for the CMM, at least for forward evaluation, so it may be best to leave this issue to the profile creator. The problems related to inverse evaluation of discontinuous curves will be discussed below.

It is worth pointing out that arbitrary parameter values can lead to strange curve shapes. Discontinuities can be large enough that the function values go out of bounds and get clipped. If the discontinuities are negative, monotonicity can be dramatically violated in such cases as well. (Figure 26.2 above is an example of this behavior.)

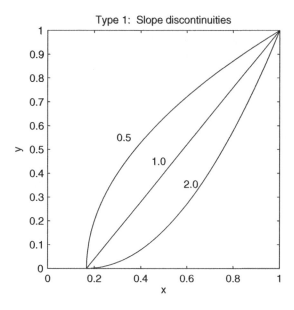

Figure 26.12 $f_1(x)$, showing effect of γ on smoothness

26.5 Smoothness

In most cases, it is not enough for the curve to be continuous at the segment boundary: it should also be smooth. This means that the first derivative must be continuous across the boundary.

In the case of types 1 and 2, the first derivative is zero in the lower (flat) segment. The derivative of the power law at the boundary will also be zero if $\gamma > 1$, and the function will then be smooth. On the other hand, if $\gamma = 1$, the derivative will be equal to one, and if $\gamma < 1$, it will diverge; in these cases, the curve will take an abrupt bend at the segment boundary. These effects are evident in Figure 26.12.

For types 3 and 4, smoothness can be achieved only by satisfying the condition

$$c = a\gamma(ad+b)^{\gamma-1},$$

as well as the continuity condition discussed above.

In general, smoothness is a concern for the profile creator, not for the CMM.

26.6 Inverse Evaluation

Curves in output profile lutAToBType and lutBToAType tags do not normally need to be inverted since the inverse of the lutAToBType is provided by the lutBToAType and vice versa; and similarly for MPE tags, where both forward and inverse functions are provided. Input profiles, where the lutBToAType may not be present, are not inverted in normal practice.

Matrix/TRC profiles contain xTRCTag curves (where x is red, green, or blue) which in v4 profiles can be defined as parametricCurveTypes, and such profiles commonly require to be inverted by the CMM in order to map PCS values back to the data encoding.

Several kinds of problems can arise when a parametric curve needs to be evaluated in the inverse direction. The most common problem occurs when there is a finite segment of the domain in which the function is constant, or *flat*. If $y = k$ (a constant) for all x in a subinterval $[x_1, x_2]$, there is no unique inverse at k, since any value in that subinterval is a legitimate candidate. While mathematically the inverse simply does not exist, computationally we may say that the inverse is ambiguous (non-unique), and attempt to remove the ambiguity.

For instance, in type 1 the function is identically zero in the lower segment $[0, -b/a]$. For $y > 0$, the inverse is simply

$$x = F_1(y) = (1/a)\left(y^{1/\gamma} - b\right)$$

but at $y = 0$ the inverse is ambiguous: it can be any value in the range $[0, -b/a]$. Figure 26.13 shows the inverse of the curve of Figure 26.5. Some CMM developers may choose to return 0 for the inverse at $y = 0$, on the grounds that the function passes through zero at zero, and that that feature should be retained in the inverse. Others may choose to return $-b/a$, on the grounds of continuity. Still others may decide to split the difference and return $-b/(2a)$.

Flat segments are explicit in the definitions of types 1 and 2. They can also occur in types 3 and 4 if the slope parameter, c, in the lower segment has the value of zero. Furthermore, flat segments can be produced by the clipping of out-of-bounds values to zero or one.

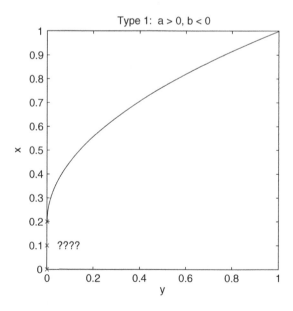

Figure 26.13 $F_1(y)$, ambiguous inverse at $y = 0$

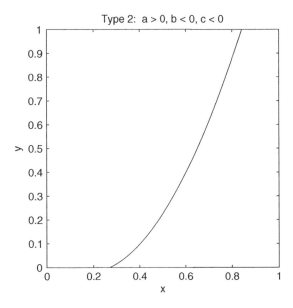

Figure 26.14 $f_2(x)$ with clipping

In a type 1 or 2 curve, if $a + b > 1$, the argument s will reach one before $x = 1$. The curve will then go out of bounds at that point, and clipping will create a flat segment with $y = 1$ at the end of the unit interval.

Further, suppose that the offset parameter c is negative in a type 2 curve. Then the function values will be clipped to zero throughout the lower segment and for some portion of the upper segment. In effect, this will create a flat segment with $y = 0$. See Figure 26.14 for an example of a curve with two flat segments, due to clipping at zero and at one.

Clipping can also affect type 3 and type 4 curves. In fact, since these curves can also be discontinuous (and non-monotonic) at the segment boundary, the functions can be clipped to zero or one in either the lower or the higher segment, or both. The "pathological" curve of Figure 26.2 is a type 4 curve with these properties.

No matter how a flat segment arises, it presents an inversion ambiguity, which must be resolved by the CMM.

Other inversion problems can occur when the curve misses some values of y in the unit interval. For instance, if c is positive in a type 2 curve, values of x in the lower segment will produce $y = c$, and values of x in the upper segment will produce values of $y > c$. No value of x will produce a value of y below c. (See Figure 26.9 for an example.) Thus, for $y < c$, the inverse is completely undefined. CMMs may well vary in their handling of this situation: some may return 0, others $-b/a$ or some other value. Similarly, in a type 1 curve, if $a + b < 1$, the argument s will never reach one (for x in the unit interval). Values of $y > (a + b)^\gamma$ will never occur, and their inverse will be ambiguous. In this case, most CMMs would return 1 for these y values. (See Figure 26.15 for a type 1 curve with missing y values at both ends of the range.)

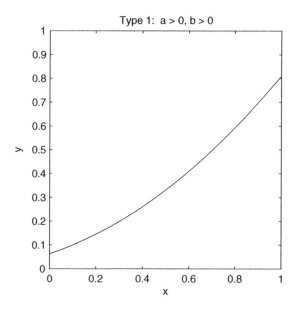

Figure 26.15 $f_1(x)$ with missing y values, both high and low

Inversion problems can also result from non-monotonicity. For instance, in a type 4 curve, a negative value of the slope parameter c will produce a decreasing function in the lower segment, followed by an increasing function in the upper segment. There can then be a range of y values that are visited twice – once by each segment of the curve. See Figure 26.16 for an example. One

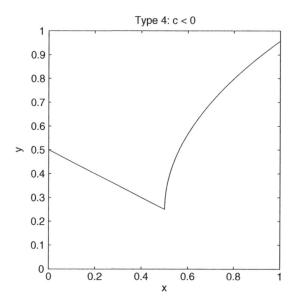

Figure 26.16 $f_4(x)$ with non-monotonicity

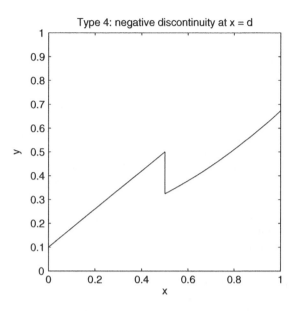

Figure 26.17 $f_4(x)$ with reversal

possible resolution is to treat negative c as an error: the CMM can then adjust c (and also the offset parameter f) so as to remove the non-monotonicity.

Non-monotonicity problems can also result from negative discontinuities, or reversals. If these are not treated as errors, there will have to be some other way to resolve the ambiguity in inversion. Figure 26.17 shows such an example.

26.7 MPE Curves

The multiProcessElementsType introduces new curve types in the ICC specification, including formula curves (defined similarly to the parametricCurveType) and sampled curve segments. The issues of imaginary numbers, continuity, and smoothness discussed above also apply to MPE segmented curves. However, continuity and smoothness may have different considerations since MPE curves are unbounded and segmented curves are most likely to be used to perform the clipping that is required to set up inputs to following CLUT elements.

26.8 Recommendations

The chief difficulty in making recommendations is the balance between the interests of the profile creators and those of the CMM developers. The functional forms designed for parametricCurveType may have originally been based on particular pre-existing use cases.

The parameter sets that support those use cases satisfy various implicit constraints and conditions.

It could be possible for the ICC to make those constraints explicit, so that CMMs could be configured so as to impose these conditions as requirements for validity and to treat violations as errors. However, it is often the case that a standard specification can be used in creative ways that were not anticipated by the original designers. Wherever possible, profile creators should be free to explore and exploit these approaches without having their efforts rejected as errors by a CMM. The constraints obeyed by the original use cases may, therefore, be too strict a guide for future profiles. From that point of view, it is better to err on the side of leniency and to impose only those constraints that are necessary to avoid the most serious problems.

It would probably be a mistake to over-analyze the situation – to enumerate all the possible combinations of parameters, to identify the interactions among them, and to formulate a complicated set of rules and exceptions to the rules. (For instance, negative s could be permitted when γ is an integer, but this is a rather uncommon – and probably pointless – exception to the general rule.) Complicated rules will be difficult to interpret and will tend to create confusion. It is better to adopt a relatively simple set of constraints, based on accepted general principles.

An important simplifying principle, in this regard, is that of *monotonicity*. If all curves were required to increase monotonically, one of the main causes of inversion ambiguity would disappear. However, it is clear that parametricCurveType has been defined so as to enable flat segments, either by design (as in types 1 and 2) or through the mechanism of clipping, so strict monotonicity is too strong a constraint.

Hence, a weaker form of monotonicity is recommended here – that the curves be *monotonically non-decreasing*. Flat segments would then be permitted (but only through the aforementioned mechanisms), and the resulting ambiguities in inversion must be dealt with. But a number of computational problems can be eliminated by adhering to this principle.

First, the condition $\gamma > 0$ can be imposed as a requirement. The power law will then be an explicitly increasing function of its argument. Furthermore, for types 1 through 4, the condition $a > 0$ can be imposed, so that the argument itself is an increasing function of x. Accordingly, all the parametric curves will be monotonically increasing wherever their behavior is determined by a power law. Flat segments will occur only through clipping or in a segment explicitly created by partitioning the domain.

If the domain is partitioned, the lower segment is defined to be flat in types 1 and 2, but it can have a non-zero slope in types 3 and 4, according to the value of c. For the sake of consistency, $c = 0$ should be allowed, but the principle of weak monotonicity rules out negative values of c. This implies the condition $c \geq 0$ for types 3 and 4.

In types 3 and 4 there is also the possibility of violating monotonicity because of a negative discontinuity (or reversal) at the segment boundary. The corresponding condition that would be imposed to prevent such reversals is

$$cd \leq (ad + b)^{\gamma}$$

for type 3 and

$$cd + f \leq (ad + b)^{\gamma} + e$$

for type 4.

In addition to monotonicity, the CMM must adhere to the *reality principle*: no imaginary or complex values. This simply means that the CMM can require that the argument of the power law be non-negative. This is an issue for types 3 and 4 and can be handled by imposing the condition

$$ad + b \geq 0.$$

Curves that violate these conditions will to some extent be "undefined" – that is, implementation dependent. A CMM could then treat such violations as errors. The response of the CMM may vary, depending on whether it is configured as an interactive tool, an operating system (OS) service, a critical process embedded in a real-time system, or something else. It might reject the profile or the entire job in which it appears, exiting with an error message. Or it might silently substitute a generic profile known to satisfy the conditions. Or it might substitute default values for the problematic parameters.

26.9 Parameter Substitutions

As an illustrative example, here is one way that the last alternative might be configured.

First, for all function types, if $\gamma \leq 0$, substitute $\gamma = 1$.
Next, for types 1, 2, 3, and 4, if $a \leq 0$, substitute $a = 1$.

For types 3 and 4, if $ad + b < 0$, substitute $d = -b/a$. Then determine whether the segment boundary is in the unit interval, that is, $0 < d < 1$. If so, check for a negative discontinuity: for type 3, if

$$cd > (ad + b)^{\gamma}$$

substitute $c = (1/d)(ad + b)^{\gamma}$ (this c will be non-negative); for type 4, if

$$cd + f > (ad + b)^{\gamma} + e$$

we may need to adjust two parameters. First, if

$$f > (ad + b)^{\gamma} + e$$

substitute

$$f = (ad + b)^{\gamma} + e.$$

Then, whether f has been modified or not, substitute

$$c = (1/d)[(ad+b)^\gamma + e - f]$$

which will also be non-negative. If the segment boundary is not in the unit interval, the function will be defined by either the lower segment (if $d \geq 1$) or the higher segment (if $d \leq 0$), but not both. If it is the lower segment (i.e., the linear expression), then the slope parameter c must still be checked for negativity; if $c < 0$, it can be reset to zero.

These procedures still allow positive discontinuities to occur at the segment boundary, as well as discontinuities of slope.

26.10 Inversion Ambiguities

Even after applying the conditions above, the CMM will still have to deal with inversion ambiguities. While CMMs may well differ in the action taken when processing an "undefined" curve, they should be consistent in their processing of valid curves, in both the forward and inverse direction. Based on (weak) monotonicity and continuity as the simplifying principles, the following procedure is recommended.

If a flat segment results from clipping to $y = 0$, the inverse at that point should be defined as the upper end point of that segment in x. If a flat segment results from clipping to $y = 1$, the inverse at that point should be defined as the lower end point of the segment. If a flat segment is defined explicitly as the lower segment ($x < d$) of types 2, 3, or 4, the inverse at that value of y should be defined as $x = d$.

An ambiguity arising from missed y values can be handled as follows. If the curve begins with a non-zero value y_0 at $x = 0$, thus skipping over $y < y_0$, the inverse for those values is defined as $x = 0$. If the curve ends at a value $y_1 < 1$ at $x = 1$, thus skipping any values of $y > y_1$, the inverse of those values is defined as $x = 1$. If there is a (positive) discontinuity at $x = d$ (for types 3 and 4), thus skipping over an interval in y, the inverse of any values in that gap is defined as $x = d$. (Negative discontinuities, as discussed above, result in "undefined" behavior and need not be considered further.)

26.11 Non-parametric Curves

Note that parametricCurveType is not the only place in the ICC specification where such issues can occur. For the sake of consistency, it is advisable to provide similar guidance for the interpretation of curveType. There are two issues here.

First, the one-dimensional LUT in a curveType tag may have a length of one. This is an exceptional case, in which the single value is interpreted as γ in a simple power law. It is encoded as an unsigned u8Fixed8Number, so negative values cannot occur. However, γ could be zero, and that case should be considered as "undefined" for consistency with the treatment of the parametricCurveType outlined above.

Secondly (and more seriously), there can be inversion ambiguities in the curveType. The one-dimensional LUT can have flat segments, and some rule is needed for defining the inverse at these points. The rule should be consistent with the one given in Section 26.10 above, but it needs to handle a greater variety of cases: flat segments in the LUT can occur anywhere, not just at the beginning or at the end of the unit interval. In addition to flat segments, reversals (violations of weak monotonicity) can also occur in the LUT; in such cases, the inverse should be designated as "undefined." Missing y values can be handled in a manner similar to the rules described in Section 26.10 above.

27

Embedding and Referencing ICC Profiles

To ensure that the color data in an image or document is correctly interpreted, many file formats permit an ICC profile defining the image source or destination to be embedded into the file or referenced by means of a uniform resource identifier (URI). Profile embedding has the advantage that the profile is permanently associated with the file and this association will not be lost during subsequent image processing or file management operations.

ICC profiles can be embedded or referenced in a wide range of file formats. This chapter summarizes the mechanical details of such operations, and provides pointers to sources of more detailed or comprehensive information.

The profile header incorporates a flags field that contains flags indicating whether a profile exists as an independent file or has been embedded into an image or document file. This flag is intended to provide hints for the CMM for purposes such as distributed processing and caching. A one in bit position 0 indicates the profile is embedded, and a further one in bit position 1 indicates that the profile cannot be used independently of the color data of the image file it is embedded within.

Embedding a profile does not guarantee that it will be used when the image is processed, as this will depend on whether the application is color management aware and on any color management run-time settings which may affect processing choices.

Image file formats that permit profile embedding fall into two types: those that define the procedure for profile embedding within the file format specification, and those that do not. In the latter case, the ICC provides the necessary information in Annex B of the ICC specification. This applies to PICT, EPS, TIF, JFIF (JPEG), and GIF formats.

All types of ICC profile except Abstract and DeviceLink profiles can be embedded into image files. When a profile is embedded, the complete file must be embedded without modification.

Color Management: Understanding and Using ICC Profiles   Edited by Phil Green
© 2010 John Wiley & Sons, Ltd

27.1 Embedding Profiles in EPS, TIF, and JPEG Files

Annex B of the ICC specification describes how an ICC profile can be associated with an EPS preview or page description. Within the page description, a %%BeginICCProfile: comment is used to mark the start of an embedded profile, with an %%EndICCProfile comment used to mark the end of the profile. Each line of profile data then begins with a % sign followed by a space, so that the profile is treated as a PostScript language comment.

An ICC profile is embedded in a TIF file as a private tag, as described in detail in Annex B of the ICC specification. The profile is stored as an image file directory (IFD) entry in the IFD that contains the image data. The IFD entry defines the tag identifying the tagged element as an ICC profile, together with the size and byte offset of the profile data.

The JPEG standard ISO/IEC 10918-1 [1] allows an ICC profile to be stored as one or more application-specific data segments, using the APP2 marker and beginning the profile data with the byte sequence "ICC_PROFILE." Large profiles cannot be stored in a single segment, so are stored as a sequence of chunks.

JPEG files also support the specification of an image color space using the Exchangeable Image File Format (EXIF) color space tag. The profiles themselves are not stored, the color space tag providing an indication of suitable profiles to use in interpreting the image data. Currently only sRGB and Adobe RGB (1998) are defined for this tag, but since multiple profiles can exist for a single color space, there can be some ambiguity in the use of the tag. When a JPEG file is opened, if the reader does not decode the tag in which the profile is stored correctly the profile is likely to be damaged or removed when the file is resaved.

Details of embedding ICC profiles in JPEG2000 files are given in ISO/IEC 15444-2 [2] and Colyer and Clark [3].

27.2 Embedding Profiles in DNG Files

The Digital Negative (DNG) file format, defined by Adobe Systems [4], provides for the embedding of an ICC profile in two locations: an AsShotICCProfile (for profiles embedded by the camera manufacturer) and a CurrentICCProfile (for profiles embedded by a raw file editor). Both profiles can be used in conjunction with a matrix which is applied before the profile.

The DNG format provides an extensive range of color calibration tags, which are intended to specify the conversion from the sensor data stored in the DNG file and a scene-referred colorimetric color space. The ICC profile additionally provides any tone and gamut mapping required to convert the scene-referred image data to an output-referred color encoding.

27.3 Embedding and Referencing Profiles in PDF Documents

Methods of including ICC profiles in PDF files are described in ISO 32000 [5].

In PDF Version 1.3 and above, ICCBased color space is included in the CIEBased color space family. This enables an ICC Device or ColorSpace profile to be embedded to define the source color space of an object in a PDF document. Each object within the PDF file can be associated with an ICC profile in this way.

Except when used in compositing, only the AToBx transform is used to interpret the source colors. The rendering intent specified in the ICC profile is ignored since the rendering intent for the document is specified elsewhere in the PDF file.

An ICCBased color space is specified within the PDF file as an array: [/ICCBased stream]. The stream requires entries defining the number of color components and an alternate color space (needed only when the PDF consumer may not be able to interpret the profile, such as when the ICC version of the profile is higher than that supported by the PDF version), followed by the embedded profile itself. PDF Version 1.7 supports ICC v4, while earlier PDF Versions 1.3–1.6 support various versions of ICC v2.

An example of an ICCBased color space embedding an ICC profile is given in 8.6.5.5 of ISO 32000.

PDF Versions 1.4 and above also support inclusion of output intents. These are included for the document rather than for individual graphic objects, and define the intended destination color space of the file. For the output intent subtype used in PDF/X files [6], an OutputIntent dictionary includes:

- the OutputCondition string;
- the OutputConditionIdentifier string, which should normally correspond to a printing condition registered in the ICC Characterization Data Registry;
- the RegistryName string, which gives the URI where the description of the registered printing condition can be found;
- an Info string providing further details of the intended output device (required if the OutputConditionIdentifier does not correspond to a registered printing condition);
- the DestOutputProfile, which consists of an ICC output profile encoded as described for the ICCBased color space. The DestOutputProfile is only required if the OutputCondition-Identifier does not correspond to a registered printing condition.

Examples of output intent dictionaries are given in 14.11.5 of ISO 32000.

PDF/X files have a single OutputIntent, in which the S key of the OutputIntent array is set to GTS_PDFX. PDF files conforming to PDF/X-1a and above should include either the name of a registered printing condition in the OutputConditionIdentifier key or an output profile in the DestOutputProfile key.

PDF files conforming to PDF/X-4p and PDF-5pg include a DestOutputProfileRef key instead of a DestOutputprofile key. The DestOutputProfileRef key includes the profile description, one or more URIs from which the profile can be downloaded, a 16-byte MD5 hash value for the profile, and the profile data color space. Full details are given in the parts of ISO 15930 corresponding to the PDF/X version.

27.4 OpenXPS

OpenXPS is the XML-based document format, originally based on the Microsoft XML Paper Specification, now undergoing standardization by the European Computer Manufacturers Association (ECMA). Each color object in an OpenXPS document has a source color space, defined as sRGB, scRGB, or an ICC profile associated with the object. As well as RGB, gray, CMYK, N-channel and Named Color spaces are supported.

Images which are part of an OpenXPS document can have an embedded profile, using the method of embedding defined by the image file format specification. Images and vector objects can also have an associated ICC profile, which may be identified as an external resource through a URI.

When associating a profile with an image the syntax is

```
<ImageBrush ImageSource=
  "{ColorConvertedBitmap
  ../Resources/Images/image.tif
  ../Metadata/profile.icc}" .../>
```

where `../Resources/Images/image.tif` is the URI of the ImageSource and `../Metadata/profile.icc` is the URI of the profile.

More details of embedding ICC profiles in OpenXPS documents is given in the Open XML Paper Specification, available from ECMA.

27.5 Interpreting Images with Embedded Profiles in HTML Documents

Most current browsers support embedded profiles in both v2 and v4 formats, the principal exception being Internet Explorer. The extent of such support varies between browsers: in one case color management must be explicitly enabled by the user, while in another images in color spaces other than RGB are supported.

Cascading Style Sheets (CSS) [7] are used in conjunction with HTML and XML to provide descriptions of how pages should be rendered. At the time of writing, the proposed CSS3 defines a color property via HTML and SVG keywords and RGB hex values. In an earlier CSS3 Candidate Recommendation, a "color profile" property was defined, together with a "rendering intent" property, but these are not included in the current working draft. However, they may be included in future levels of CSS.

There are resources on the ICC web site for checking the extent to which browsers and other applications support color management, embedded profiles, and the current specification version. See http://www.color.org/version4ready.html for details.

References

[1] ISO/IEC (1994) 10918-1:1994. *Information technology – Digital compression and coding of continuous-tone still images: rrequirements and guidelines.* International Organization for Standardization, Geneva.
[2] ISO/IEC (2004) 15444-2:2004. *Information technology – JPEG 2000 image coding system: extensions.* International Organization for Standardization, Geneva.
[3] Colyer, G. and Clark, R. (2003) *Guide to the Practical Implementation of JPEG 2000*, PD 6777:2003. British Standards Institute, London.
[4] Adobe Systems (2008) Digital Negative (DNG) Specification Version 1.2.0.0.
[5] ISO (2008) 32000-1:2008. *Document management – Portable Document Format – Part 1: PDF 1.7.* International Organization for Standardization, Geneva.
[6] ISO (various dates) 15930. *Graphic technology – Prepress digital data exchange using PDF – Parts 1–8.* International Organization for Standardization, Geneva.
[7] W3C CSS (2008) Color Module Level 3, W3C Working Draft, 2008.

28

LUT-Based Transforms in ICC Profiles

LUT-based transforms in ICC profiles are used to define a mapping between two color encodings. They can be used when the transform cannot be defined through the curve and matrix elements provided in the matrix/TRC type of profile. Table look-up and interpolation usually has a lower computational cost when the function to be encoded cannot be expressed algebraically in a simple linear form.

It should be emphasized that the ICC specification does not prescribe the content of transforms in an ICC profile. In order to assure interoperability, it specifies the format and interpretation of the processing elements, and imposes some basic requirements for handling the dynamic range and gamut of the medium.

In an LUT-based transform a sample of known outputs of a function is stored, allowing unknown values to be found by interpolation. This differs from an algebraically defined function which can be applied directly to the input data, since for most inputs the value will not be in the input table and must be interpolated.

The ICC profile defines a lutAToBType and a lutBToAType in Versions 4.0 and above. A simpler LUT type, originally defined in earlier versions of the specification but still supported within the specification, can also be used. Many of the features of the two types are common, so here the v4 type will be described and a summary of differences between v2 and v4 types given later.

A v4 lutAToBType or lutBToAType can include a PCS-side curve, a data encoding-side curve, a matrix and a matrix curve, as well as the multi-dimensional color LUT (CLUT). The complete set of elements and possible combinations are shown in Figures 2–3 of the ICC specification.

Since the multi-dimensional CLUT will usually be the main basis of the transform, this is described first and the use of the other elements in conjunction with the CLUT (or alone where the CLUT is not present or is an identity transform) is described later.

The construction and use of a CLUT involves a number of discrete operations. For a given input table, the output values are computed and encoded in the required form, and then stored in the profile. This process is referred to as packing or partitioning [1]. At run-time, for each color to be transformed to the output encoding a subset of values from the table is extracted and then used to calculate the output value by interpolation.

Some of the methods for these operations are covered by patents, and Kang [1] identifies certain patents in this area. Neither this chapter nor the ICC specification can provide a complete listing of patents in this field.

The CLUT packing and encoding operations are completed during the profile generation stage, and are independent of the methods used for extraction and interpolation. As a result the encoded CLUT should be designed to give similar results regardless of the methods used for extraction and interpolation.

Characterization models for determining the relationship between device and PCS encodings in order to perform the packing operation are described elsewhere [1–6]. The profile creator should use a modeling technique which is suited to the physical characteristics of the particular medium.

A CLUT is a means of encoding a function, where "function" is taken to mean a relation that associates the values in an input domain to those in an output range, such that each element in the set of values of the domain is associated with a value in the output encoding. While many functions can be expressed algebraically, a CLUT provides an alternative, non-algebraic means of defining the relation between the two sets. The set of input values is limited by the domain, which in ICC profiles is normalized to [0, 1]. The set of output values associates a single output value with each input; there is no limit on the number of values in the input domain that can be associated with a single value in the output range.

A CLUT can be thought of as an association of an input table with an output table. In ICC profiles, the input table is always normalized in the range [0, 1] and is always uniformly spaced, so that it is completely determined by the number of "channels" (color components) and the number of stored values (or "nodes") in each dimension. Since the ICC LUT-based types store the parameters needed to reconstruct the input table, the input table entries themselves are redundant and are not stored in the profile.

The number of input channels i and output channels o are stored in the eighth and ninth bytes respectively of the lutAToBType and lutBToAType. The offset to the beginning of the CLUT is stored as a uint32Number, and the size of the CLUT in bytes is $(nGrid1 \times nGrid2 \times \cdots \times nGridN) \times o \times p$, where $nGrid1$ is the number of lattice entries in the first input channel, $nGrid2$ is the number of lattice entries in the second input channel, $nGridN$ is the number of lattice entries in the Nth input channel, and p is the precision of the entries (1 or 2 bytes). If the number of lattice entries is the same for all input channels, the size of the CLUT is also given by $n^i \times o \times p$, where n is the number of grid entries in each dimension.

An input table can be represented in column form, with one column per channel. The entries in the table are uniformly spaced, with the rightmost column varying most rapidly and the leftmost column varying least rapidly. Since the table is normalized to [0, 1], the eight bounding nodes are represented by the input table below. An input table with three nodes or lattice points per channel is shown in Figure 28.1.

0	0	0
0	0	1
0	1	0
0	1	1
1	0	0
1	0	1
1	1	0
1	1	1

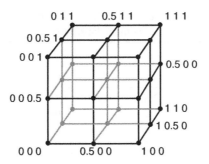

Figure 28.1 Grid or lattice for CLUT transform between three-component spaces, with three lattice points per color component

If there are four input components, the additional components can be added as shown below. Figure 28.2 shows a lattice with four input components and three lattice points per component.

```
0   0   0   0
0   0   0   1
0   0   1   0
0   0   1   1
0   1   0   0
0   1   0   1
0   1   1   0
0   1   1   1
1   0   0   0
1   0   0   1
1   0   1   0
1   0   1   1
1   1   0   0
1   1   0   1
1   1   1   0
1   1   1   1
```

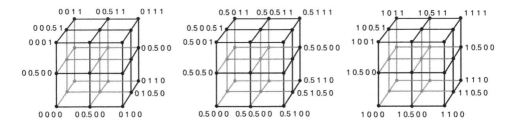

Figure 28.2 Grid or lattice for CLUT transform with four input components

Table 28.1 Example measurement values for a CLUT with three dimensions and two nodes per dimension, in CIELAB L^*, a^*, b^*, together with media-relative scaled values

D50 measurements			Media-relative PCSLAB		
L^*	a^*	b^*	L^*	a^*	b^*
6.57	0.12	−2.65	7.35	−0.28	−1.89
14.93	8.19	−27.2	16.04	8.36	−26.54
20.22	26.83	−47.66	21.54	27.6	−47.1
21.83	−26.97	−0.51	23.23	−28.31	0.9
23.63	−18.59	−20.32	25.1	−19.53	−19.16
35.18	4.84	−50.83	37.07	4.55	−49.74
31.32	−49.89	8.76	33.07	−51.95	10.62
36.71	−52.83	−7.94	38.65	−55.13	−6.12
50.26	−34.31	−44.51	52.7	−36.09	−42.77
18.01	14.34	−1.98	19.26	14.53	−0.76
19.16	21.57	−15.81	20.44	22.17	−14.78
28.91	37.26	−42.13	30.56	38.21	−41.14
28.56	−9.02	18.1	30.21	−9.77	20.02
35.83	−10.03	−4.4	37.74	−10.81	−2.55
46.73	11.11	−43.49	49.04	10.99	−41.86
47.64	−54.34	40.92	49.98	−56.71	43.9
55.91	−58.82	12.93	58.56	−61.44	15.78
68.89	−35.58	−34.72	72.03	−37.53	−32.14
46.05	66.91	49.39	48.33	68.73	52.46
48.46	64.89	25.56	50.84	66.61	28.34
55.05	66.29	−16.41	57.67	67.99	−14.05
58.81	43.33	64.91	61.57	44.2	68.67
64.76	38.58	28.69	67.74	39.23	32.11
69.88	42.95	−18.32	73.04	43.74	−15.45
84.7	−1	86.34	88.41	−1.88	91.4
90.01	−3.35	48.8	93.92	−4.38	53.47
95.87	0.91	−4.05	100	0	0

Additional nodes can be added to each dimension as required, depending on the precision required. For example, for a CLUT with three dimensions and three nodes in each channel, the normalized input table (shown in Figure 28.1) would be

```
0.0  0.0  0.0
0.0  0.0  0.5
0.0  0.0  1.0
0.0  0.5  0.0
0.0  0.5  0.5
0.0  0.5  1.0
0.0  1.0  0.0
0.0  1.0  0.5
0.0  1.0  1.0
0.5  0.0  0.0
```

```
0.5   0.0   0.5
0.5   0.0   1.0
0.5   0.5   0.0
0.5   0.5   0.5
0.5   0.5   1.0
0.5   1.0   0.0
0.5   1.0   0.5
0.5   1.0   1.0
1.0   0.0   0.0
1.0   0.0   0.5
1.0   0.0   1.0
1.0   0.5   0.0
1.0   0.5   0.5
1.0   0.5   1.0
1.0   1.0   0.0
1.0   1.0   0.5
1.0   1.0   1.0
```

This input table samples the normalized input domain uniformly, with equal numbers of nodes in each channel. This uniform spacing is a requirement of CLUTs in v2 lut8 and lut16 types, but v4 lutAToBType and lutBtoAType CLUTs are permitted to have a different number of nodes in each channel. The first 16 bytes of the lutAToBType and lutBtoAType CLUT encodings are used to specify the number of nodes in each channel.

Having generated the table that samples the input domain, we can now obtain output values. In the example in Table 28.1, D50 measurements of the coordinates in the input table are shown. These are converted to media-relative values by first transforming to XYZ, then scaling using Equation 24.1 in Chapter 24, and finally converting back to CIELAB.

The CIELAB values in Table 28.1 are encoded as described in the ICC specification. The L^* values are scaled by multiplying by 255/100 or 65 535/100 (depending on whether the table is to be encoded with 8- or 16-bit precision), while the a^* and b^* values are given an offset of 128 and then either encoded directly as uint8Numbers or multiplied by 257 to encode as uint16Numbers. For the media-relative values in Table 28.1, this results in the encoded values in Table 28.2.

The encoded values in Table 28.2 are in the form required to be stored in the profile. They are written in interleaved order so that for three-component data the first three entries (8 or 16 bits according to the chosen precision) correspond to the first node in the table, the first row of Table 28.2.

In the case of an AToBx CLUT (where x is 0, 1, or 2, according to whether the rendering intent is perceptual, colorimetric, or saturation), the whole of the domain can be sampled and output values measured or estimated.

In a BToAx CLUT, the input table is a uniform sampling of the PCS domain. The CLUT domain can be made non-uniform with respect to the PCS by a nonlinear B curve, as described in Chapter 25. For all realizable device gamuts, most of the values in the PCS encoding will lie outside the gamut of the device.

The CLUT nodes in the input table are fixed and it is not practical to physically realize these exact values, even for the in-gamut colors, on the input or output medium. In consequence, the CLUT output values for the BToAx direction in the device encoding are estimated using a

Table 28.2 The 8- and 16-bit encodings for the media-relative output values in Table 28.1

8-bit PCSLAB encoding values			16-bit PCSLAB encoding values		
PCSLAB L	PCSLAB a	PCSLAB b	PCSLAB L	PCSLAB a	PCSLAB b
19	128	126	4817	32824	32410
41	136	101	10512	35045	26075
55	156	81	14116	39989	20791
59	100	129	15224	25620	33127
64	108	109	16449	27877	27972
95	133	78	24294	34065	20113
84	76	139	21672	19545	35625
99	73	122	25329	18728	31323
134	92	85	34537	23621	21904
49	143	127	12622	36630	32701
52	150	113	13395	38594	29098
78	166	87	20027	42716	22323
77	118	148	19798	30385	38041
96	117	125	24733	30118	32241
125	139	86	32138	35720	22138
127	71	172	32754	18322	44178
149	67	144	38377	17106	36951
184	90	96	47205	23251	24636
123	197	180	31673	50560	46378
130	195	156	33318	50015	40179
147	196	114	37794	50369	29285
157	172	197	40350	44255	50544
173	167	160	44393	42978	41148
186	172	113	47867	44137	28925
225	126	219	57939	32413	56386
239	124	181	61550	31770	46638
255	128	128	65535	32896	32896

characterization model (combined with matrix and curve elements as required). In practice this is also commonly the case for the AToBx CLUT, since it may not be feasible to sample all the coordinates that are used in the input table. Furthermore, using a model to estimate the nodes, rather than basing the output table on direct measurements, leads to a smoother underlying function in the output table.

The procedure for populating the BToAx table is similar to that for the AToBx table described above, with the additional step of gamut mapping to provide in-gamut output values for all the values in the CIELAB encoding domain. Gamut mapping techniques are described elsewhere; in an ICC profile the choices are essentially clipping (for colorimetric intents), compression (for perceptual and saturation intents in BToA0 and BToA2 tables) where the output gamut is almost invariably smaller than the PRMG, and gamut expansion in AToB0 and AToB2 tables where the input gamut is smaller than the PRMG. The gamut mapping technique used in the perceptual and saturation intents should ideally compress or expand the gamut as required, depending on the relative gamut size.

Although the PRMG was defined after the publication of the ICC v2 and the v4.1 and 4.2 specifications, it is nevertheless a requirement for all v4 profiles to use a reasonable reference medium, and ambiguities in the interpretation of v2 profiles can similarly be avoided by the use of a reference medium gamut (RMG). In v4 profiles the "rig" tag can be used to indicate the use of the PRMG in generating the profile.

28.1 Matrix

In lutBToAType and lutAToBType transforms, the matrix is 3×4 and enables a linear transform from PCS values before the CLUT (in the case of lutBToAType tags) or a linear transform from CLUT output values prior to the PCS. For example,

$$y = Mx$$

where x and y are 3×1 and M is 3×3.

If the fourth column of the matrix contains non-zero values a constant offset is also applied and the transform is affine. This can also be expressed as a matrix multiplication:

$$\begin{bmatrix} y \\ 1 \end{bmatrix} = \begin{bmatrix} M & b \\ 0,0,0 & 1 \end{bmatrix} \begin{bmatrix} x \\ 1 \end{bmatrix}$$

in which M is the first three columns of the matrix and b is the fourth column.

The most common use of a matrix is to convert between the XYZ PCS and an RGB encoding that is a linear transform of XYZ, such as sRGB. The linear transform is expressed in matrix form and the matrix coefficients and offsets are converted to s15Fixed16Number types.

Note that the range of the matrix is likely to be different from its domain, and if the range is to be normalized or otherwise modified this must be done in the curve following the matrix.

In v2 profiles, the matrix is 3×3 and is required to be the identity unless the PCS is XYZ. More details on the use of matrices in v4 profiles can be found in Chapter 29.

28.2 Curves

The curves are used in conjunction with the CLUT to provide a transform. The ICC specification does not define precisely how they should be used, as this is up to profile creators. One common use of the curves is to ensure that the CLUT itself is optimized. For example, if the data encoding for a printer is nonlinear with respect to the PCS, a curve can be used to make it more linear and thus distribute the look-up and interpolation errors more uniformly throughout the space. Curves can also be used to give greater emphasis to certain parts of the color space (usually neutrals) or to reduce the number of CLUT grid points used for coordinates that are outside the medium gamut.

More details on the use of curves in lutAToBType and lutBToAType transforms is provided in Chapter 25.

28.3 Combining Matrix, Curve, and CLUT Elements

The possible combinations of these elements in a profile are as follows:

lutAToBType

B curves
M curves – matrix – B curves
A curves – CLUT – B curves
A curves – CLUT – M curves – matrix – B curves.

lutBToAType

B curves
B curves – matrix – M curves
B curves – CLUT – A curves
B curves – matrix – M curves – CLUT – A curves.

Where an element is not used in the transform, its tag offset is set to zero. Where an element is required to be present according to the list above, but is not used in the transform, it can be set to an identity transform.

This flexibility in the combination over transform elements makes the ICC LUT structures both powerful and flexible. The profile builder can optimize the transform for the requirements of the application, and the resulting structure provides an unambiguous transform operating on colorimetry that can be relative to the appropriate adopted white. The adopted white is either the medium or a perfectly reflecting diffuser, and in the case of media-relative rendering intent a well-defined color gamut is associated with the transform so that an output profile does not require knowledge of the input medium or its profile or color gamut.

The steps in creating a lutBToAType tag with A and B curves can be summarized as follows:

1. Select the precision of curves and CLUT elements.
2. Select the domain and range of the CLUT that will optimize the transform.
3. Determine the transforms to be applied by B curves that will produce values in the domain of the CLUT.
4. Normalize the output values of the transform to the range [0, 1].
5. Encode each of the B curves as a curveType tag or a parametricCurveType tag.
6. Generate the input table for the CLUT.
7. Select the number of grid points in each dimension of the CLUT.
8. Compute the media-relative output values for each input entry.
9. Encode the CLUT values as uint8Number or uint16Number entries.
10. Determine the transforms to be applied by the A curves that will operate on the range of the CLUT, and normalize as above.
11. Encode each of the A curves as a curveType or a parametricCurveType tag
12. Encode the type signature, reserved bytes, number of input and output channels, and the offsets to each of the elements in the first 32 bytes of the lutBToAType.

The matrix and M curves can be added if used. A similar set of steps is performed to create an lutAToBType tag.

Some examples of combining matrix, curve, and CLUT elements are given below.

1. Convert PCSXYZ to CMYK printer, using colorimetric density as the domain of the CLUT

   Colorimetric density is the base 10 log of the tristimulus value [4]. It is a more perceptually uniform domain than XYZ, and is better correlated with CMYK colorant amounts than CIELAB. In printer characterization it is usually applied in media-relative form, such that

$$
\begin{bmatrix} D_r \\ D_g \\ D_b \end{bmatrix} = \begin{bmatrix} \log 10 \left(\dfrac{X_0}{X} \right) \\ \log 10 \left(\dfrac{Y_0}{Y} \right) \\ \log 10 \left(\dfrac{Z_0}{Z} \right) \end{bmatrix}.
$$

   The PCS values in the lutBToATypes are already normalized to the media white, so the conversion from PCSXYZ to colorimetric density can be performed by applying a base 10 log in the B curve. The matrix and M curves are not used.

   The final one-dimensional A curves are then used as a look-up between the CLUT output and the fractional colorant amount, and will usually be close to linear. In this case the relationship between colorimetric density and colorant amount might be found by measurement of single color ramps: the curves could then be constructed by a cubic spline interpolation of the measured values.

   Alternatively the A curves can be used to adjust for the tone response of the printer in a calibration step.

   The CLUT converts between the colorimetric density of the domain to the linearized range of the fractional colorant amount. This CLUT will tend to be more perceptually uniform and the visual impact of interpolation errors is thus likely to be minimized.

2. Convert PCSXYZ to RGB printer, using perceptually linear RGB as the domain of the CLUT

   The matrix and M curve are used to convert PCSXYZ to a more perceptually linear RGB. The A curves represent a nonlinear transform between perceptually linear RGB and printer RGB, and can be found by regression. The CLUT then converts perceptually linear RGB to linearized printer RGB.

3. Convert PCSLAB to printer CMYK or RGB, using neutral-weighted CIELAB as the domain of the CLUT

   By making the B curves for the a^* and b^* channels of PCSLAB sigmoidal in shape, they will compress the domain in the flatter parts of the curve and give greater emphasis to the central region corresponding to the neutral part of the PCSLAB encoding. The A curves are found by regression between this neutral-weighted PCSXYZ and the RGB or CMYK of the device encoding, as for example 1 above, and the CLUT converts between the neutral-weighted CIELAB domain and the linearized device encoding.

Similar considerations apply to creating the transforms for lutAToBType tags.

28.4 Inverting the LUT-Based Transform

For most applications the BToAx and AToBx transforms in a profile should invert accurately, so that when an image is roundtripped (convert first in one direction and then back in the other), all the final in-gamut colors are a close approximation to the starting values, allowing for minor differences arising from interpolation errors and round-off. This allows an AToBx transform to be used to preview the effect of a conversion using a BToAx tag. Simulation, re-rendering, and re-purposing all require the BToAx and AToBx tags to invert accurately.

Below we will assume that an AToBx CLUT has been generated from a uniform sampling of the device encoding and the associated measurements, and we will consider the methods of generating its inverse, the BToAx CLUT.

In the simplest case, the function used to calculate the AToBx output table is linear and analytically invertible. This might be the case for a 3×3 matrix, for example. When this inverse function is used to compute the inverse transform, the results will be sufficiently accurate for most purposes. Alternatively, for three-component data the output values might be computed by polynomial regression and, while the resulting coefficients do not lend themselves to a simple inversion, the regression can be recalculated with the sense inverted.

Other functions cannot be inverted so readily. An alternative method is to find the direct inverse of the CLUT, as described by Bala [2]. The goal of the method is to use the mapping in one direction to compute the mapping in the inverse direction. The input cube is partitioned into tetrahedra, and the output table is similarly tessellated such that each tetrahedron in the input table is mapped uniquely to an output tetrahedron. Any point in the output encoding can now be mapped to the input encoding by a process of locating it within a tetrahedron in the output encoding and using tetrahedral interpolation to find its value in the input encoding. All the values in a uniformly spaced lattice in the output encoding are converted in this way, and the resulting values represent the output of the inverse CLUT.

Any values that lie outside the tetrahedra representing the medium gamut must of course be mapped to one of the tetrahedra in order to invert it. There are no restrictions on the method used to partition the input lattice into tetrahedra.

In most cases the curves and matrix in a profile can readily be inverted. This is discussed further, including special cases where a curve or matrix cannot be inverted, in Chapter 26. A 3×4 matrix in an lutAToBType and lutBToAType transform is inverted through a 3×1 offset which contains the values of the fourth column of the forward matrix with a change of sign, followed by the matrix inverse of the first three columns of the forward matrix. This inverse of an affine transform can also be expressed in matrix form as

$$\begin{bmatrix} M^{-1} & -M^{-1}b \\ 0,0,0 & 1 \end{bmatrix}.$$

28.5 Version 2 LUTs

Many profiles continue to be generated according to the ICC v2 specification. While the v2 lut8Type and lut16Type LUTs lack the flexibility of the v4 lutAToBType and lutBToAType, there is no problem in using them in most workflows unless v4 profiles are required for some reason. Profiles using these v2 LUTs will interoperate with v4 profiles and current CMMs.

A v2 LUT type lacks the pre-matrix curve and must have a constant number of grid points in each dimension of the table. Unlike v4 LUTs, the type signature denotes the precision rather than the direction of the transform. The type signature is thus "mft1" or "mft2" for 8- or 16-bit tables respectively, rather than "mAB" or "mBA" for v4 AToBx and BToAx LUTs respectively.

Matrices are 3×3 instead of 3×4, with no provision for a constant offset as in a v4 LUT type. The matrix is required to be the identity when the PCS is not PCSXYZ, and the use of the matrix is thus confined to linear transforms from XYZ prior to applying the input table and CLUT. It should be noted that the PCSXYZ encoding is 16 bit only, and so lut8Type profiles using PCSXYZ as the PCS should be avoided as at best the results will be implementation specific.

28.6 Transform Quality

When evaluating a transform, the accuracy with which it returns values in the desired output function will normally be the primary consideration. However, other aspects of the transform will affect the quality of the profile in its intended application.

Of particular importance will be the smoothness of the output function and its performance in neutrals and in high-chroma colors close to the gamut boundary.

Transforms should be tested with different CMMs, software applications, and test images. CMMs differ in the methods of extraction and interpolation used, and some CLUT packings may perform well in one interpolation scheme and badly in another. If possible, tests should be performed with CMMs using both trilinear and tetrahedral interpolation schemes.

Test images should include both natural and synthetic images. Synthetic images can be designed to evaluate smoothness and gamut boundary performance, while natural images are important to evaluate the transform performance with real images.

Similarly, a range of applications should be tested, including as many as possible of those that will be used in the intended application.

References

[1] Kang, H.R. (2006) *Computational Color Technology*, SPIE Press, Bellingham, WA.
[2] Bala, R. (2003) Device characterization, in *Digital Color Imaging* (ed. G. Sharma), CRC Press, Boca Raton, FL.
[3] Bala, R. and Klassen, V. (2003) Efficient color transformation implementation, in *Digital Color Imaging* (ed. G. Sharma), CRC Press, Boca Raton, FL.
[4] Green, P.J. (2002) Overview of characterization methods, in *Colour Engineering* (eds P.J. Green and L.W. McDonald), John Wiley & Sons, Ltd, Chichester.
[5] Green, P.J. (2002) Characterizing hard copy printers, in *Colour Engineering* (eds P.J. Green and L.W. McDonald), John Wiley & Sons, Ltd, Chichester.
[6] Green, P.J. (2006) Accuracy of color transforms. *Proc. SPIE*, **6058**, Article 2.

29

Populating the Matrix Entries in lutAtoBType and lutBtoAType of Version 4 ICC Profiles

29.1 Introduction

One of the improvements of the Version 4 ICC profile specification is the addition of a set of constants to the matrix operation in lutAtoBType and lutBtoAType. The specification describes a simple matrix operation; however, there are some subtleties to using the matrix properly. This chapter describes how to populate the entries of the matrix to achieve the expected results.

29.2 The Matrix Operation

The specification describes 12 coefficients, e_1 to e_{12}, which are used in a matrix operation. Using x_1, x_2, and x_3 as inputs and y_1, y_2, and y_3 as outputs, the operation is

$$y_1 = x_1 e_1 + x_2 e_2 + x_3 e_3 + e_{10}$$

$$y_2 = x_1 e_4 + x_2 e_5 + x_3 e_6 + e_{11} \qquad (29.1)$$

$$y_3 = x_1 e_7 + x_2 e_8 + x_3 e_9 + e_{12}.$$

The inputs and outputs are defined to be values in the range 0.0–1.0. There are several encodings (e.g., those used for PCSLAB and PCSXYZ) that may be used as inputs and outputs, but when the CMM implements the matrix operation it does not perform any further conversions based upon the specific encodings. It simply applies the above operations to the inputs and passes the outputs to the next processing element defined in the profile.

Color Management: Understanding and Using ICC Profiles   Edited by Phil Green
© 2010 John Wiley & Sons, Ltd

29.3 The Matrix Entries

The coefficients required for a given operation will depend on the equations to be used, the range of the inputs, and the range of the outputs. The matrix operates on inputs in the range [0.0, 1.0]; therefore, the range of the inputs must be mapped to the input range of the matrix. This mapping is substituted into the equations for the operation. Similarly, the outputs from the matrix are in the range [0.0, 1.0] and need to be mapped to the range of the outputs. This mapping is also substituted into the equations for the operation. Note that these formulas are used to help determine the coefficient values and are *not* implemented by the CMM!

In general, mapping ranges is done by mapping the minimum of the first range to the minimum of the second range and the maximum of the first range to the maximum of the second range. Intermediate values are linearly spaced between these minimum and maximum values. The equation is

$$y = (x - \min_1)(\max_2 - \min_2)/(\max_1 - \min_1) + \min_2 \qquad (29.2)$$

where:

x is a value in range 1;
y is a value in range 2;
$\min_1$ and $\max_1$ are the minimum and maximum of range 1;
$\min_2$ and $\max_2$ are the minimum and maximum of range 2.

The matrix operates on values mapped as shown, so it is necessary to know the formulas for converting mapped input values to input values and mapped output values to output values. On the input side, the input to the matrix uses the range [0.0, 1.0], and so $\min 1 = 0.0$ and $\max 1 = 1.0$. The input mapping formula in Equation (29.2) then becomes

$$iv = miv(\max_2 - \min_2) + \min_2 \qquad (29.3)$$

where iv is an input value and miv is the mapped input value.

The output from the matrix is in the range [0.0, 1.0], the same as for the input. Just as for the input mapping the output mapping formula becomes

$$ov = mov(\max_2 - \min_2) + \min_2 \qquad (29.4)$$

where ov is the output value and mov is the mapped output value resulting from the matrix operation.

Once the mapping formulas have been substituted into the operation equations, the input variables are separated to determine the coefficients. Each of these coefficient values is converted to an s15Fixed16Number to obtain the final value to be written into the matrix.

29.4 Example

In this example we determine the matrix coefficients for performing the linear part of the PCS Lab-to-XYZ conversion. It is assumed that the curveType preceding the matrix has an identity operation.

Inverting the equations in Annex A of the specification gives

$$fX = a/500 + (L+16)/116$$
$$fY = (L+16)/116 \qquad (29.5)$$
$$fZ = -b/200 + (L+16)/116.$$

The matrix operates on mapped values, so the mapping formulas must be determined. The preceding curveTypes implement an identity operation, so the inputs to the matrix operation will have PCS Lab encodings. For L^*, the range is 0.0 to 100.0, so $max_2 = 100.0$ and $min_2 = 0.0$. Using $m_L =$ mapped L^*,

$$L = m_L(100.0 - 0.0) + 0.0$$
$$= m_L 100. \qquad (29.6)$$

The PCS a^* and b^* ranges are -128 to $+127$, so $max_2 = 127$ and $min_2 = -128$. Using $m_a =$ mapped a and $m_b =$ mapped b,

$$a = m_a(127 - (-128) + (-128))$$
$$= m_a 255 - 128$$
$$b = m_b(127 - (-128) + (-128)) \qquad (29.7)^{**}$$
$$= m_b 255 - 128.$$

Substituting these equations into the operation equations gives

$$fX = (m_a 255 - 128)/500 + ((m_L 100) + 16)/116$$
$$fY = ((m_L 100) + 16)/116 \qquad (29.8)$$
$$fZ = -(m_b 255 - 128)/200 + ((m_L 100) + 16)/116.$$

fX, fY, and fZ are in the range 0.0–1.0 and do not correspond to any particular encoding. They may be left in this state, or some additional function may be applied to them. This is a choice left to the profile builder. Whichever choice is made, it will need to be coordinated with the calculation of the curveType that follows the matrix. In this example, no additional function is applied, so no output mapping equations need to be applied to the equations.

Separating variables gives

$$fX = (m_a 255 - 128)/500 + ((m_L 100) + 16)/116$$
$$= m_L(100/116) + m_a(255/500) + (-128/500 + 16/116) \qquad (29.9)$$

$$fY = ((m_L 100) + 16)/116$$
$$= m_L(100/116) + (16/116) \tag{29.10}$$

$$fZ = -(m_b 255 - 128)/200 + ((mL100) + 16)/116$$
$$= m_L(100/116) + m_b(-255/200) + (128/200 + 16/116). \tag{29.11}$$

Referring back to the matrix equations, $x_1 = m_L$, $x_2 = m_a$, $x_3 = m_b$, $y_1 = fX$, $y_2 = fY$, and $y_3 = fZ$. Substituting into the above equations gives

$$y_1 = x_1(100/116) + x_2(255/500) + (-128/500 + 16/116)$$
$$y_2 = x_1(100/116) + (16/116) \tag{29.12}$$
$$y_3 = x_1(100/116) + x_3(-255/200) + (128/200 + 16/116).$$

From this we see that the matrix coefficients are

$$e_1 = 100/116$$
$$e_2 = 255/500$$
$$e_3 = 0$$
$$e_4 = 100/116$$
$$e_5 = 0$$
$$e_6 = 0$$
$$e_7 = 100/116 \tag{29.13}$$
$$e_8 = 0$$
$$e_9 = -255/200$$
$$e_{10} = -128/500 + 16/116$$
$$e_{11} = 16/116$$
$$e_{12} = 128/200 + 16/116.$$

These are all s15Fixed16Numbers in the profile format. Their encodings are generated by multiplying by 65 536, rounding to the nearest integer, and converting to hexadecimal. The actual numbers in the profile are

$$e_1 = DCB1h$$

$$e_2 = 828Fh$$

$$e_3 = 0h$$

$$e_4 = DCB1h$$

$$e_5 = 0h$$

$$e_6 = 0h$$

$$e_7 = DCB1h$$

$$e_8 = 0h$$

$$e_9 = FFFEB99Ah$$

$$e_{10} = FFFFE1C6h$$

$$e_{11} = 234Fh$$

$$e_{12} = C726h.$$

$$(29.14)$$

30

Implementation Notes for SampleICC's IccProfLib

The SampleICC project is an open source object-oriented C++ development effort that was written to provide an example of how various aspects of ICC color management can be implemented. The basic SampleICC was originally written by Max Derhak as part of an MS degree in Imaging Science at Rochester Institute of Technology. After extensive revisions suggested by the ICC Architecture Working Group, the project was assigned to the ICC as a means of helping to describe the approaches to color management implementation implied by the ICC color profile specification. Within the ICC Architecture Working Group this library has evolved to allow for prototype development and refinement of internal proposals before they become part of the ICC specification. The library now allows extensions to be added without requiring changes to the core libraries, and can be downloaded from http://sampleicc. sourceforge.net.

The SampleICC project contains a platform-independent library (named IccProfLib) that provides a complete implementation for reading, writing, and applying ICC profiles. The IccProfLib subproject has HTML documentation that describes the classes and their interfaces, but the basic relationship between the classes as it relates to applying profiles is not necessarily clear. This chapter complements the IccProfLib class documentation by describing how the objects interact when applying profiles. Some familiarity with both object-oriented programming and the ICC profile specification is assumed, and overview information will be given related to classes within IccProfLib. For specific details, readers should consult the implementation as defined by the source code. It should be noted that IccProfLib was initially named IccLib, but has since been changed to avoid conflicts with existing libraries.

There are many different ways to implement color management. In Chapter 6 general classes of color management are outlined. The implementation presented here represents the fulfillment of a "static" CMM with most of the "smart" color rendering operations contained in the ICC profiles themselves. Both a fixed and a programmable CMM model (based upon multi-processing elements) are possible within the IccProfLib framework.

Color Management: Understanding and Using ICC Profiles   Edited by Phil Green
© 2010 John Wiley & Sons, Ltd

Note that this does not preclude the possibility of implementing a "dynamic" (late binding) CMM based upon the profile file support provided by IccProfLib; those wishing to participate in such an endeavor are encouraged to join the project.

30.1 Profiles, Tags, and Processing Elements

ICC color management is made possible through the use of ICC profiles. These profiles are defined by the ICC profile specification. The CIccProfile class in IccProfLib provides the basic implementation foundation for reading, writing, and otherwise manipulating the contents of ICC profiles based upon the ICC profile specification. Each CIccProfile object maintains a list of tag objects, all derived from the CIccTag class. It is the CIccTag-derived objects that provide the implementation for reading, writing, and manipulating tag data in an ICC profile.

When parsing an ICC profile the CIccProfile object first reads the ICC header and tag directory. The tag directory managed by the CIccProfile object associates 4-byte signature values with each tag as defined by the ICC profile specification. When parsing tag data from a profile the protected member function CIccProfile::**LoadTag**() first reads the tag-type signature and then uses the static CIccTag::**Create**() member function to create the tag. The created CIccTag-based object is then used to parse the contents of the tag.

The static CIccTag::**Create**() member function uses a singleton object that implements a factory design pattern to create tags. The singleton object manages a list of IIccTagFactory classes that provide the implementation for actually creating all CIccTag-based objects in the system. The CIccTag::**Create**() calls the static CIccTagCreator::**CreateTag**() function to create all tags. When this function is called, it calls each IIccTagFactory's CreateTag() function until a factory creates the tag type.

This mechanism allows for new tag types for parsing, writing, display, and validation to be easily added without having to modify any code in IccProfLib. All that is needed is an implementation of a CIccTag-derived object class and an object that implements the IIccTagFactory interface to be passed to the CIccTagCreator::PushFactory() member function.

This approach has been found to be a very useful mechanism by the ICC Architecture Working Group for prototyping and implementation of new tag types in separate implementation libraries without having to modify the base IccProfLib. Once these new tag types and changes are approved by the ICC, they can then be moved to the base IccProfLib.

In the ICC Version 4.3 profile specification there is a new tag type that allows for the direct encoding of floating point data in an arbitrary sequence of processing elements. Support for the multiProcessElementsType (MPET) tags is provided in IccProfLib through the CIccTagMPE class. The implementation of processing elements is parallel to the implementation of tags. Just as all tag classes are derived from the CIccTag class, all individual processing element classes are derived from the CIccMPE class. Similar to tags, processing elements are created using the extendible CIccElemCreateor factory design pattern class, which allows for user extension of private processing element types without modification of IccProfLib. This has proven useful as the ICC Architecture Working Group has considered possible additional processing element types. However, it is important to note that the use of private element types in a profile will result in CMMs using the profile's integer-based tag types that do not support the private element types.

30.2 CMMs in SampleICC

The following discussion makes use of Figure 30.1 to help explain how profiles are essentially applied within IccProfLib. The figure shows some basic relationships between object types used by the library to apply profiles. To provide for the ability to simultaneously apply pixels in multiple threads, this basic architecture has been extended, and the extensions will be discussed after these basic relationships are described.

ICC profile files are read, written, and otherwise manipulated through the use of CIccProfile objects that have attached CIccTag objects which contain the associated profile tag data. The application of profiles is implemented separately through the use of a CIccCmm class object.

A CIccCmm class object is used to administer and perform color management transforms. This object manages a list of CIccXform-derived objects which are associated with corresponding CIccProfile objects. Each CIccXform object obtains information from its corresponding CIccProfile object and attached CIccTag objects in order to perform the requested color transformations. The basic relationships between objects in IccProfLib are shown in Figure 30.1.

Profile application is performed by using the CIccCmm::**Apply**() method after a CIccCmm object has been properly constructed and initialized. This method makes use of the list of CIccXform objects with their associated CIccProfile objects to perform color space transforms.

When applying pixel colors, the CIccCmm object assumes that data pixels have been converted to an internal floating point encoding with all values ranging from 0.0 to 1.0. Results from calling the CIccCmm::**Apply**() method are also in this range. The CIccCmm class provides overloaded conversion member methods (CIccCmm::**ToInternalEncoding**() and CIccCmm::**FromInternalEncoding**()) to facilitate conversion to/from typical encoding range values used by various pixel formats. The CIccCmm class provides the static Boolean member function CIccCmm::**IsInGamut**() to determine whether the internal representation of the result is in gamut when using a gamut tag from a profile.

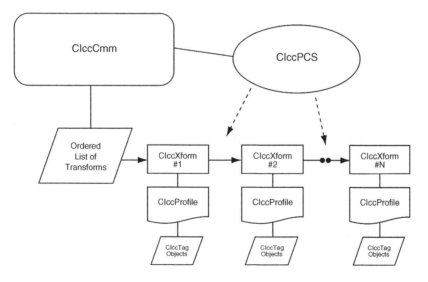

Figure 30.1   Basic object relationships in IccProfLib

30.3 CIccCmm Details

There are five stages in the life of a CIccCmm object:

1. **Creation.** The following information is provided to a CIccCmm object when it is created:
   (a) The source and destination color spaces are identified. In many cases the color spaces can be specified as undefined – the color spaces will be determined by the attached profiles.
   (b) In preparation for the next stage, the initial transform side (input vs. output) is identified.
2. **Attachment.** One or more calls to a CIccCmm::**AddXform**() method are used to attach one or more ICC profiles to the CIccCmm object. There are several overloaded versions of this method:
   (a) In one version the first argument is the file path to the ICC profile file.
   (b) In another overloaded version, the first argument is a pointer to the CIccProfile object with the profile already loaded. The ownership of the CIccProfile object is passed to the CIccCmm object.
   (c) In another overloaded version, a reference is passed to a CIccProfile object. The CIccProfile object is copied with the copy owned by the CIccCmm Object.
   (d) In another version the first argument is a pointer to a memory-based ICC profile file, with the second argument being the length of the file in memory.
   Regardless of which version is used, the CIccCmm object keeps track of whether the input or output side of an attached profile should be used. It also keeps track of the connecting color spaces and ensures compatibility. Any number of profiles can be attached to a CIccCmm object. The order in which profiles are attached to the CIccCmm object defines the order in which the appropriate transforms will be applied.
   Only a single transform from a profile will be used for each attached profile. Since multiple transforms (in separate tags) can be stored in a single ICC profile, CIccCmm:: **AddXform**() arguments are used to determine which transform should be used. The nIntent argument allows selection between rendering intents, and the nLutType argument allows selection between the color, preview, and gamut tags. NamedColor profile selection is automatically detected when using the basic color transforms. Here it should be noted that the preview and gamut transforms are considered to be output transforms for automatic input/output transform selection purposes.
   CIccCmm::**AddXform**() creates a CIccXform object for each profile as it is attached both to keep track of the profile and to provide the implementation of the transformation using data from the profile.
3. **Initialization.** This is done in two parts:
   (a) The CIccCmm::**Begin**() method is used to indicate that no more profiles will be attached, and that color transformation processing will now begin. The CIccCmm:: **Begin**() method performs final color space verification and then each attached CIccXform object is initialized (using the CIccXform::**Begin**() method) to begin color transformations.
   (b) Data members of the CIccCmm and CIccXform objects that may be modified during the apply process are allocated and initialized.

4. **Apply.** A CIccCmm::**Apply()** method can be used to apply the ordered sequence of CIccXform objects to source pixel(s) to arrive at destination pixel values. This method uses a CIccPCS object with the initial color space to keep track of the current color space as transforms are applied. A temporary pixel is also defined and modified within the method. The source pixel, temporary pixel, and destination pixel are all involved in the concept of the "current" pixel. For each transform in the ordered CIccXform object list the "current" pixel is checked with the CIccPCS::**CheckPCS()** method to make sure that the current color space agrees with the input color space of the next transform. The adjusted pixel is then passed to the CIccXform::**Apply()** method to perform the pixel transformation. Once the last transform is performed, the CIccPCS::**CheckLast()** method is used to make any final color space adjustments.

5. **Destruction.** The CIccXform list and its accompanying objects are released.

IccProfLib provides support for all color profile types (ICC.1:2009–XX Sections 9.3 through 9.9). All color profile types except named color profiles (ICC.1:2009–XX Section 9.9) are supported by the CIccCmm class. The CIccNamedCmm class (also defined in IccCmm.cpp) is derived from the CIccCmm class and supports the use of named profiles in addition to the capabilities offered by the CIccCmm class. This was done to avoid the cost of the extra overhead of supporting named colors in the basic CMM class as defined by CIccCmm. The CIccNamedCmm class provides additional CIccNamedCmm::**Apply()** method interfaces to support the input and/or output of color names. This approach allows multiple named color profiles to be linked together using color names as a connection space.

30.4 CIccXform Details

CIccXform is the base class that defines the basic interface for performing pixel transformations. There are multiple classes that are derived from this class that provide specific implementations. There are three important static member methods for the CIccXform base class:

1. The static member function CIccXform::**Create()** is used in conjuction with the CIcc-CreateXform singleton to create actual instances of CIccXform objects. This function uses a CIccProfile object argument to decide which specific CIccXform object to create. The type of CIccXform depends upon the type of transform that is implied by the ICC profile. Four types are possible: matrix/TRC, multi-dimensional look-up table, multi-processing element based, and named color indexing. The CIccXform choices include:
   (a) CIccXformMatrixTRC – Uses the RGB chromaticities and transfer functions to perform pixel transforms.
   (b) CIccXformMonochrome – Uses the gray tone reproduction curve to perform pixel transformations on single channel data.
   (c) CIccXform3DLut – Performs pixel transformation on three-dimensional input data. The extracted tag from the attached CIccProfile is determined by the rendering intent and input/output flag. The CIccXform3DLut object is also configured to perform either linear or tetrahedral interpolation.

(d) CIccXform4DLut – Performs pixel transformation on four-dimensional input data. The extracted tag from the attached CIccProfile is determined by the rendering intent and input/output flag. The CIccXform4DLut object only performs linear interpolation.

(e) CIccXformNDLut – Performs pixel transformation on *N*-dimensional input data. The extracted tag from the attached CIccProfile is determined by the rendering intent and input/output flag. The CIccXformNDLut object only performs linear interpolation.

(f) CIccXformMPE – Performs pixel transformation on input data using an ordered sequence of processing elements stored in a multiProcessElement type tag. The extracted tag from the attached CIccProfile is determined by the rendering intent and input/output flag. Each processing element provides instructions on how to apply portions of the transform

(g) CIccXformNamedColor – Performs color transforms using text strings to define the color. The static CIccXform::**Create()** method is passed an argument that specifies whether or not the calling CIccCmm object supports named colors. If named colors are not supported, then this object type will not be created.

2. The protected member method CIccXform::**CheckSrcAbs()** is called by the derived CIccXform::**Apply()** methods to perform any required absolute-to-relative colorimetry transformation. This method also handles legacy PCS encoding, and Version 4 to Version 2 perceptual black point translation. If the source color space is not a PCS color space, then this method makes no adjustments to the pixel.

3. The protected member method CIccXform::**CheckDstAbs()** is called by the derived CIccXform::**Apply()** methods to perform any required relative-to-absolute colorimetry transformation. This method also handles legacy PCS encoding, and Version 2 to Version 4 perceptual black point translation. If the destination color space is not a PCS color space, this method makes no adjustments to the pixel.

There are two virtual methods that all derived CIccXform objects need to implement:

1. The virtual CIccXform::**Begin()** method is called during CIccCmm::**Init()** to allow the CIccXform-derived object to initialize itself relative to the attached color spaces, input/output transform flag, and rendering intent. Additional important methods that are also used include:

(a) CIccXformMatrixTRC – The CIccXformMatrixTRC::**Begin()** method calculates a matrix and one-dimensional LUTs to use. In some cases an inverse matrix and LUTs are calculated. Further details can be found in the implementation.

(b) CIccXformMonochrome – The CIccXformMonochrome::**Begin()** method extracts the appropriate curve to use. In some cases an inverse curve is calculated. Further details can be found in the implementation.

(c) CIccXform3DLut, CIccXform4DLut, CIccXformNDLut – Extracts appropriate curve and LUT tags from the profile and prepares for pixel transformations.

(d) CIccXformNamedColor – Identifies the correct CIccXformNamedColor::**Apply()** interface to use based upon attached color spaces.

2. A virtual CIccXform::**Apply()** method does most of the work of color transformation. Each derived object provides the implementation of this method to perform the specific

operations that are required to implement the color transformation. The order of the operations depends upon whether the CIccXform object represents an input transformation or an output transformation. The operations by transform type are as follows:

(a) CIccXformMatrixTRC – If the CIccXform object represents an input transform the following steps are performed:
  (i) CIccXform::**CheckSrcAbs()**.
  (ii) Apply one-dimensional curves look-up.
  (iii) Apply matrix.
  (iv) CIccXform::**CheckDstAbs()**.
  If the Xform represents an output transform, the following steps are performed:
  (i) CIccXform::**CheckSrcAbs()**.
  (ii) Apply matrix.
  (iii) Apply one-dimensional curves look-up.
  (iv) CIccXform::**CheckDstAbs()**.

(b) CIccXformMonochrome – The following steps are performed:
  (i) CIccXform::**CheckSrcAbs()**.
  (ii) Apply one-dimensional curve.
  (iii) CIccXform::**CheckDstAbs()**.

(c) CIccXform3DLut, CIccXform4DLut, CIccXformNDLut – The following lists show the order of operations. Not all profile tags provide data to perform operations, in which case steps associated with missing data are simply ignored.
  If the CIccXform object represents an input transform the following steps are performed:
  (i) CIccXform:: **CheckSrcAbs()**.
  (ii) Apply one-dimensional B curves look-up.
  (iii) Apply matrix.
  (iv) Apply one-dimensional M curves look-up.
  (v) Perform multi-dimensional interpolation.
  (vi) Apply one-dimensional A curves look-up.
  (vii) CIccXform:: **CheckDstAbs()**.
  If the Xform represents an output transform, the following steps are performed:
  (i) CIccXform:: **CheckSrcAbs()**.
  (ii) Apply one-dimensional A curves look-up.
  (iii) Perform multi-dimensional interpolation.
  (iv) Apply one-dimensional M curves look-up.
  (v) Apply matrix.
  (vi) Apply one-dimensional B curves look-up.
  (vii) CIccXform:: **CheckDstAbs()**.

(d) CIccXformMPE – The following lists show the order of operations.
  If the CIccXform object represents an input transform the following steps are performed:
  (i) Apply all processing elements in sequence.
  (ii) If Colorspace is PCSXYZ or PCSLab convert from actual values to internal encoding.
  (iii) If the tag rendering intent is not absolute colorimetric call CIccXform:: **CheckDstAbs()**.

If the Xform represents an output transform, the following steps are performed:
   (i) If the tag rendering intent is not absolute colorimetric call CIccXform::**CheckSrcAbs()**.
   (ii) If Colorspace is PCSXYZ or PCSLab convert from internal encoding to actual values.
   (iii) Apply all processing elements in sequence.
(e) CIccXformNamedColor –This object type uses the CIccTagNamedColor2 tag object of the associated CIccProfile to perform the color transformations. The CIccXform-NamedColor object behaves differently than the other CIccXform object types. Different CIccXformNamedColor::**Appy()** interfaces are supported to allow for transforms involving named colors. It requires that a named color is always used as either the input or the output side of the transform. Thus direct transforms to/from device coordinates from/to PCS values are not directly supported. To accomplish this, simply attach the named profile to a CIccCmm object twice – the first time with the named color as the output, and the second time with the named color as the input. This results in two CIccXformNamedColor objects being used.

   If the input color space is a named color space the operations are as follows:
   (i) Search for color name in the named color tag.
   (ii) If the output color space is PCS then set pixel to corresponding PCS value and apply CIccXform::**CheckSrcAbs()**.
   (iii) Else (the output color space is a device color space) set pixel to corresponding device values.
   If the output color space is a named color space the operations are as follows:
   (i) If the input color space is PCS then call CIccXform::**CheckSrcAbs()** and then find the color index of the color whose PCS value has the least ΔE difference to the source color.
   (ii) Else (the input color space is a device color space) find the color index of the color whose device coordinate has the smallest Euclidean distance to the source color.
   (iii) Set destination color to the corresponding index color name.

Two points to note here are:

1. A similar mechanism is used for creating CIccXform objects that exist for creating CIccTags. Registering an IIccXformFactory object with the singleton CIccXformCreator object allows one to replace the default construction of existing transform classes without modifying the base IccProfLib library.
2. Implementing new Xform types, in conjunction with private tag types, will require a custom CIccCmm-based class to be implemented that understands and is able to connect the needed CIccTag-based objects to the created CIccXform-based objects.

30.5 CIccPCS Details

The CIccPCS object is an object that is used to keep track of the current color space as transformations are applied within the CIccCmm::**Apply()** method. In addition to storing the current color space, this object also performs necessary PCS conversions when connecting profiles with different PCS characteristics. Such differences include CIEXYZ, CIELab Legacy, and CIELab Version 4 encodings. Since each of these "color spaces" is considered to be a PCS, the CIccPCS object is used to seamlessly translate between these PCSs as needed. Two main methods are provided, in addition to color space access methods:

1. CIccPCS::**Check**() – This method checks to see if the current color space defined by the pixel is compatible with the source color space of the CIccXform object that will be using the pixel. It only makes conversions if the current color space is a PCS color space.
2. CIccPCS::**CheckLast**() – This method checks to see if the current color space defined by the pixel is compatible with the destination color space. It only makes conversions if the current color space is a PCS color space.

30.6 Extensions for Thread Safety

In initial implementations of IccProfLib the CIccCmm, CIccPCS, and CIccXform objects contained run-time data (such as the concept of the "current pixel") that would be manipulated and changed during the process of applying transforms to pixels. With this approach it was impossible for multiple threads to simultaneously apply transforms associated with a single CIccCmm without the possibility of data corruption. To overcome this restriction the member portions of these objects that could be modified were moved into separate shadow "Apply" objects that need to be created and used to perform pixel transform application.

Thus the relationships depicted in Figure 30.1 above can be replaced by a more correct version in Figure 30.2. Here the CIccCmm amd CIccXform objects are shadowed by associated Apply objects, and each "Apply" object is linked directly to its corresponding

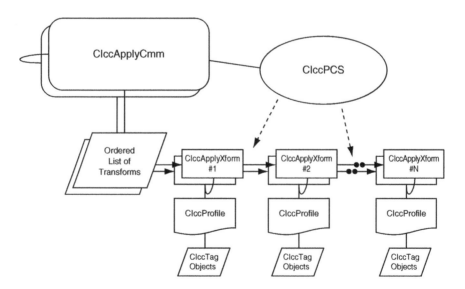

Figure 30.2   Extended object relationships in IccProfLib

CIccCmm or CIccXform object. Ownership of a CIccPCS object has been transferred to the CIccApplyCmm object.

The implementation changes necessary for this involved a change to parts 3b and 4 of the CIccCmm details section discussed earlier. In part 3b a call to CIccCmm::**GetApply**() needs to be performed to allocate and initialize data members that will be modified during subsequent calls to Apply(). In part 4 of the implementation, the Apply() function was moved to the CIccApplyCmm and CIccApplyXform objects.

The CIccApplyCmm object returned by CIccCmm::**GetApply**() can be directly used to apply pixel transformations. More than one call to CIccCmm::**GetApply**() can be made to allocate multiple CIccApplyCmm objects that can be used for simultaneous apply operations without the need for synchronization operations.

For compatibility of source code with previous implementations, there is an additional optional parameter to the CIccCmm::**Begin**() function to call CIccCmm::**GetApply**() and associate a CIccApplyCMM object with the CIccCmm object. Calls to the CIccCmm::**Apply**() member functions now call the Apply() function of this associated CIccApplyCmm object.

30.7 Implementation Details of the CIccMruCmm

The CIccMruCmm is an example of an extension CIccCmm-based class that implements a decorator design pattern to allow a CIccCmm to have basic caching of the most recently used pixels. In the implementation of the CIccMruCmm class the static member function CIccMruCmm::**Attach**() allocates and associates a CIccMruCMM with a previously initialized CIccCmm(). The CIccApplyMruCmm::**Apply**() member function first looks in a cache of recently applied pixels before calling the attached CIccApplyCmm's Apply() function.

The level of improvement of the CIccMruCmm is somewhat limited, but it does show an excellent example of implementing further, more robust, performance improvements using SampleICC's IccProfLib.

31

Introducing the New multiProcessingElements Tag Type

In November 2006 the ICC approved the multiProcessingElements tag type as part of the Floating Point Encoding Range addendum to the ICC specification. The primary purposes of this new tag type were to overcome limited precision in ICC profiles by optionally allowing for the direct encoding of floating point data in an ICC profile, remove bounding restrictions for both device-side and PCS encoding ranges, and provide for backwards compatibility with the existing ICC profile specification. A secondary benefit of this new tag type is that it provides for greater flexibility in encoding transforms.

It should be noted here that the use of floating point computation in a CMM to apply profiles does not necessarily require the encoding of floating point data in profiles.

The initial motivation for the Floating Point Device Encoding Range amendment stems from an attempt to use ICC color profiles for managing colors in motion picture and digital photography workflows. The precision of profile elements such as LUTs, curves, and matrices is insufficient for the motion picture industry processing because:

1. Encoding is limited to 16 bits and therefore transform inversion cannot be performed precisely due to quantization errors. An example of such quantization errors can be found in Figure 31.1, where a DPX scene curve is encoded in an ICC profile.

   Some of the values of the curve are shown in Table 31.1.

   The curve in this example has 1024 points, one point corresponding to each input 10-bit DPX count. The values to be encoded in the ICC profile are those in the "Normalized values" column. Since ICC profiles only support up to 16-bit precision, the values must be converted to 16 bits. It is clear from the column "16-bit values" that doing so results in severe quantization. This quantization becomes obvious when the curve is graphed on a log scale, as shown in Figure 31.2.

2. Current profile transforms only support bounded device-side color encodings, but unbounded (floating point) encodings are used in the motion picture industry.

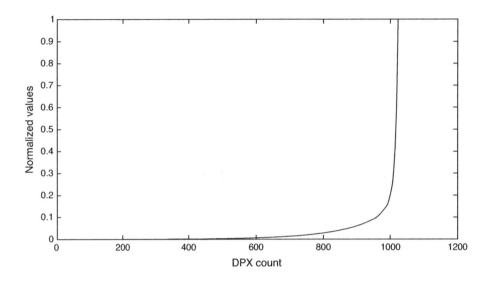

Figure 31.1 DPX scene curve

3. The PCS encoding is limited to an encoding range of [0, 2) for XYZ values or [0, 100] for L^* and [−128, 127] for a^* and b^*.

Previously there was some confusion regarding ICC support of encoding above-white values. Such values can be supported by setting the media white point appropriately and using the absolute rendering intent. Clarifying language is provided in the CIIS (Colorimetric Intent Image State) amendment and in version 4.3 and above of the specification.

Table 31.1 Values in the DPX scene curve

10-bit DPX count	Curve values	Normalized values	16-bit values
0	0.001 855	2.44594E-05	2
1	0.001 876	2.47420E-05	2
2	0.001 898	2.50279E-05	2
⋮	⋮	⋮	⋮
37	0.002 837	3.74154E-05	2
38	0.002 870	3.78477E-05	2
39	0.002 903	3.82851E-05	3
40	0.002 937	3.87274E-05	3
⋮	⋮	⋮	⋮
684	0.895 955	0.011 815 8	774
685	0.902 699	0.011 904 8	780
686	0.909 493	0.011 994 4	786
⋮	⋮	⋮	⋮
1021	58.629 983	0.773 211 9	50 672
1022	66.676 150	0.879 324 7	57 627
1023	75.826 544	1.000 000 0	65 535

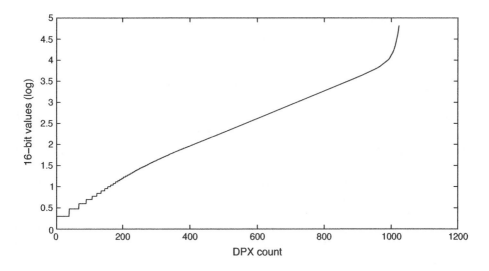

Figure 31.2 DPX scene curve, log scale showing quantization errors

Eight new tags that use this new tag type are defined to allow for the optional substitution of the tags that define rendering intents in an ICC profile. D2Bx/B2Dx tags can now exist in a profile in addition to AToBx/B2Ax tags. A CMM can now optionally first look for DToBx/BToDx tags and use them instead of the AToBx/B2Ax tags or matrix/TRC tags. Explicit definition of the absolute rendering intent is also now possible through the use of DToB3 and BToD3 tags.

The eight new tags are optional, which means that existing color management systems can ignore them as private tags. This allows for profiles to be built and embedded in images, or otherwise used in workflows that do not support the use of these new tags without breaking those workflows. It is hoped that the use and adoption of these tags will be made easier because their presence should not break existing workflows.

It should be emphasized that CMM support for the multiProcessingElements tag type is optional. This means that MPE-based tag support is not guaranteed to be provided and implemented by CMMs in general. Additionally, all required tags must be present and valid.

The DToBx/BToDx tags all make use of the new multiProcessingElements tag type. Important aspects of this tag type include the following:

1. This tag type provides for an arbitrary sequence of processing elements to perform the device-to-PCS, PCS-to-device, or device-to-device conversion. Processing elements can be thought of as transformation steps that convert input channel data to output channel data.

   The ability to have an arbitrary sequence is a significant difference from LutAToB and LutBToA tag types. Additionally, with an arbitrary sequence of elements, a "limited" static programmable CMM is implied.
2. All the processing elements encode data using 32-bit IEEE 754 floating point encoding.
3. The absolute rendering intent can be encoded with DToB3/BToD3 tags.
4. The initial repertoire of processing elements includes N-dimensional LUTs, $N \times M$ matrices, sets of one-dimensional segmented curves, and two future expansion elements that perform no operation.

5. The PCS for DToBx/BToDx tags is the floating point equivalent of the PCS in AToBx/ BToAx tags. When DToBx tags are connected to BToDx tags no clipping is performed in the PCS.

6. DToBx tags can be connected to BToAx or matrix/TRC-based tags with appropriate clipping of the PCS values as needed, and AToBx or matrix/TRC-based tags can be connected to BToDx tags.

7. The CMM performs *no* manipulation of data between processing elements. The CMM simply passes the results from one processing element to the next processing element.

8. Generally, up to 65 535 channels of floating point data can be passed between processing elements:

   (a) Processing element types are not required to support the upper limit of 65 535 input and 65 535 output channels.

   (b) The channel usage of the first and last elements in a DToBx/BToDx tag must agree with the channel usage requirements of both the containing DToBx/BToDx tag and the profile header. (Note that, currently, the color fields of the profile header limit the maximum number of channels to 15.)

9. The fallback behavior of using AToBx/BToAx tags is prescribed if processing elements are encountered that are unknown to the CMM. This allows for a graceful handling for future expansion of the processing element repertoire.

10. The device encoding range for DToBx and BToDx tags is unbounded, but conversions and clipping may need to be made to be compatible with AToBx and BToAx tags. The equivalent device encoding range of AToBx and BToAx tags is converted to the range of 0.0–1.0 when applying DToBx and BToDx tags.

Figure 31.3 shows possible scenarios using DToBx/BToDx profiles. In the figure:

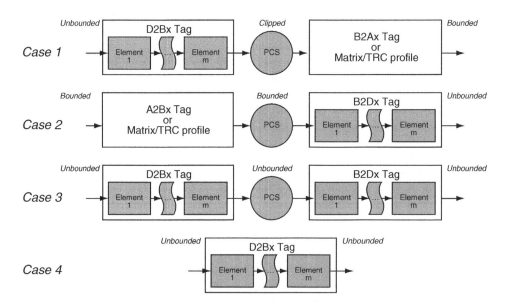

Figure 31.3   Example D2Bx/B2Dx scenarios

- Case 1 shows the connection of a profile using a DToBx tag to a profile containing a BToAx tag or a matrix/TRC-based profile.
- Case 2 shows the connection of a profile containing a AToBx tag or a matrix/TRC-based profile to a profile using a BToDx tag.
- Case 3 shows the connection of a profile containing a DToBx tag to a profile containing a BToDx tag.
- Case 4 shows the use of a DToBx tag as a device link or an abstract profile.

31.1 multiProcessingElementsType Overview

The multiProcessingElementsType generally encodes the following:

1. A tag type signature ("mpet") followed by a reserved 32-bit integer set to zero.
2. The number of input and output channels associated with the tag. These should match the number of channels associated with color space fields in the profile header.
3. The number of elements in the transform along with an ordered array of processing element position entries. The element position order defines the sequence of element processing. Each position entry contains an offset relative to the tag start as well as the size of the element data. Multiple position entries can refer to the same data.
4. Data for all processing elements associated with the tag.

31.2 Processing Elements

The initial repertoire of processing elements includes N-dimensional LUTs, $N \times M$ matrices, sets of one-dimensional segmented curves, and two future expansion elements that perform no operation. These all use 32-bit IEEE 754 floating point encoding for processing and data storage purposes. An arbitrary number of these elements can be combined in any order to accomplish the purpose of defining a transform. Output channels from preceding elements are direct inputs to succeeding elements.

The color LUT or CLUT element (with signature "clut") is used to store N-dimensional LUTs. They can accept up to 16 input channels (constrained by the allowed "Number of grid points in each dimension" in the CLUT element encoding) and output up to 65 535 channels. The CLUT input range is from 0.0 to 1.0 since using grid points represents the sampled range, and clipping is prescribed for values outside this range. Scaling/conversion may need to be performed by a processing element before a CLUT element to get values into the range from 0.0 to 1.0. The output range of a CLUT element is the entire floating point encoding range.

The matrix element (with signature "matf") can be used to store an $N \times M$ matrix with a constant offset vector. The input and output dimensions need not be the same, and up to 65 535 channels can be used for both input and output. The input and output range is the entire floating point encoding range. (Note that this is different from other LUT tag types that can only store and use 3×3 matrices with offset.)

The curve set element (with signature "cvst") encodes multiple one-dimensional curves. Up to 65 535 separate curves can be defined. The curves are segmented to allow the entire

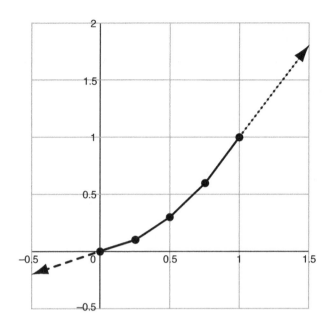

Figure 31.4   Example curve set element curve

floating point encoding range to be used as both input and output. Up to 65 535 segments are possible for each curve with the positions and definitions of each segment definable. Each segment can be defined as a formula or sampled curve segment:

- Formula segments define a function type and provide parameters to the function.
- Sampled segments are equally spaced sample points defining a one-dimensional LUT. An important aspect of sampled segments is that interpolation is performed from the last point of the previous curve segment. This means that the first interpolation point is *not* stored in the sampled segment.

To help exemplify this, Figure 31.4 illustrates an example curve set element curve. In the figure:

- The segmented curve contains two break points that define three segments with break points at $x = 0.0$ and $x = 1.0$.
- The first (lower) segment is a formula segment that has a domain of $[-\infty, 0.0]$, and is defined by the formula $y = 0.1x$.
- The second (middle) segment is a sampled segment that has the domain $(0.0, 1.0]$ with four sampled values: 0.1, 0.3, 0.6, and 1.0.
      Notice that the sampled segment does not store a value for the $(0, 0)$ position as it gets its interpolation point from the first segment.
- The third (upper) segment is a formula segment that has the domain $(1.0, +\infty]$, and is defined by the formula $y = 1.6x - 0.6$.

Possible additions to the processing element repertoire are also under consideration within the ICC. Two future expansion element types were included in the specification (with signatures "bACS" and "eACS" for expansion purposes). These elements encode a single signature value and have no prescribed operations – thus they define no operation on channel data. They simply pass the channel data to the next processing element.

31.3 Example

A DToBx tag could possibly be used to encode a device model that goes from device values to spectral information and then to PCS. A sequence of processing elements can be used to model each step of the transformation.

The following example sequence of processing elements in a hypothetical DToBx tag (i.e., the actual element boxes that would go into the DToBx tag element boxes of Figure 31.3) helps to show the programmable nature of this new tag type:

1. The first processing element is either an N-dimensional LUT element or a matrix element that converts the device channel data to 31 channel spectral information.
2. The second processing element uses a 3×31 matrix to convert spectral information to XYZ colorimetric information, as shown in Figure 31.5.

   At this point, if the PCS signature in the profile header is "XYZ" then the transformation is complete (in other words, only two elements are needed to go from device to PCS). However, if the PCS signature in the profile header is "Lab" then the next two processing elements could be included.
3. The third element uses a curve set to perform the cube-root portion of an XYZ to Lab conversion.
4. The fourth element uses a 3×3 matrix to complete the XYZ to Lab conversion, as shown in Figure 31.6.

Figure 31.5   Device-to-XYZ example

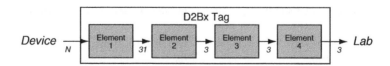

Figure 31.6   Device-to-PCSLAB example

It is important to note that the CMM has no understanding of what each of these processing elements is actually accomplishing. To the CMM this is interpreted as follows:

1. The device values are passed through either a matrix or N-dimensional LUT to get 31 values.
2. A 3×31 matrix is applied to these 31 values to get three values.
3. Separate one-dimensional curves are applied to each of these three values to get three new values.
4. A 3×3 matrix is applied to the resulting three values from step 3 to get three values that are assumed to be PCSLAB values.

Note that this example is rather basic. It is readily admitted that these elements could probably be combined, but the example does show the programmable nature of using MPE tags.

31.4 Implementations

At the time of writing, the following implementation are known:

- A reference implementation of a C++ programming library with applications that are capable of parsing, applying, and validating profiles containing D2Bx/B2Dx tags can be found in the SampleICC project (http://sourceforge.net/projects/sampleicc/). The ApplyNamedCmm and wxProfileDump application are capable of opening, displaying the contents, and applying profiles. Additionally, third-party application support can be provided through the Windows SampleICCCMM.dll.
- Adobe has released a version of the Adobe CMM that has support for MPE profiles.
- LittleCMS 2.0 (in beta at the time of publication) has implemented multiProcessing-ElementType support.

Other implementations may also exist. In addition, example ICC probe profiles containing MPE-based tags that make it possible to determine whether a CMM has support for MPE-based tags, and for which rendering intent will be applied, can be found on the ICC web site.

Constructing a multiProcessingElements Profile

A multiProcessingElements (MPE) profile has greater flexibility in defining sequences of processing elements, and uses 32-bit floating point numbers to encode curve values and parameters. Not all CMMs will process MPE DToB or BToD tags, so AToB and BToA tags are provided for interoperability with other CMMs. If these tags are used, the transform will not be an exact match to that implied by the MPE tags.

Tag	Size (bytes)	Value
"desc"	92	DPX Scene Standard Camera Film
"A2B0"	6412	v4 lutAToBType with A curves, 3D CLUT, M curves, 3 × 4 matrix, and B curves
"B2A0"	6412	v4 lutBToAType with M curves, 3 × 4 matrix, and B curves
"D2B0"	152	v4 multiProcessElementsType with matrix and curve set elements[a]
"B2BD"	152	v4 multiProcessElementsType with matrix and curve set elements
"wtpt"	20	[73.112 17, 75.826 54, 62.548 57][b]
"cprt"	110	Adobe Systems Inc.

[a] Curve set elements in this profile include formula curve segments and sampled curve segments, enabling encoding of a highly nonlinear transform. Entries in the curve segments are encoded as float32Numbers.

[b] Defines a `media white' with a luminance of 7583 cd/m^2, corresponding to a brightly lit scene.

32

Inverting ICC Profiles

The ICC specification ICC.1:2004–10 states that: "In general, the BToAxTags represent the inverse operation of the AToBxTags." This chapter is intended to provide further clarification on this statement in situations where its intention may be unclear.

In profiles where only a matrix/TRC transform is provided, the inverse is computed by the CMM. A 3×3 matrix can be inverted analytically, as can a curve represented by a scalar exponent. A curve represented by a one-dimensional table or a parametricCurveType is inverted by mapping input values to output values, interpolating as required to give new output values for a uniformly spaced input table.

A v2 or v4 LUT-type transform is normally non-invertible (except in special cases), and hence the inverse transform is explicitly encoded in the profile; for example, the inverse of an lutAToBType is encoded in the lutBToAType in the same profile.

32.1 Limits to Inversion Accuracy

For the profile classes DisplayDevice, OutputDevice, and ColorSpace, it is desirable that a conversion performed by applying the profile to data can be inverted for purposes such as previewing and proofing. Ideally, a roundtrip operation (in which the data is converted from one encoding through the profile to another and then back to the original encoding, using the same rendering intent) should result in values that exactly match the starting values. In practice, differences arise due to:

- finite precision of computation, and rounding and interpolation operations;
- the sampling of the encoding in the CLUT;
- differences in color gamut in the PCS;
- intentional differences between BToAx and AToBx transforms.

Mismatches between the number of components of the PCS and the number of data encoding channels of the profile will also give rise to different possibilities for inverses.

Color Management: Understanding and Using ICC Profiles   Edited by Phil Green
© 2010 John Wiley & Sons, Ltd

Finite precision of computation and rounding operations give rise to relatively small differences on roundtripping. These will depend largely on the CMM used to apply the profile, but the profile generator can minimize such differences by using 16-bit data in LUTs. In situations where the magnitude of roundtrip errors is too high for a particular application, higher precision floating point LUT encodings using DToBx and BToDx tags could be considered.

In a BToAxTag, output values are given at uniformly spaced intervals for the entire PCS encoding. Many of the values in the PCS will be outside the gamut of the data encoding, and the BToAxTag maps these out-of-gamut colors to coordinates within the range of the data encoding. In the AToBxTag, all input values have an output value in the PCS, and no mapping of out-of-gamut colors is required. In the colorimetric intents only the values within the gamut of the data encoding can be roundtripped accurately, while in the perceptual and saturation intents only colors within the Perceptual Reference Medium used in the profile BToAx and AToBx tags can be roundtripped accurately.

The effective device values are the set of device values that results from mapping the entire PCS to the output encoding. It is only possible to roundtrip to within this set of values. The AToBTag maps these device values to the PCS, which enables the effective gamut to be determined. The profile can only accurately roundtrip within this gamut.

For example, a given CMYK value has a unique PCS value in the AToBxTag, but when this PCS value is converted back to the data encoding with the BToAxTag, the resulting CMYK value will depend on the black generation and ink limiting choices made during profile generation and not on the starting CMYK value.

Since the colorimetric intents in a profile must be measurement based, it is implicit that the BToA1Tag is required to be an accurate inverse of the AToB1Tag. A preview or proof of colors in the data encoding is normally obtained by applying the colorimetric intent AToB1Tag, regardless of which intent was used in the BToAx direction. In most situations, the AToB0Tag and AToB2Tag should be the inverse of the BToA0Tag and BToA2Tag respectively, to allow data to be converted consistently between PCS and data encodings.

Exceptionally, a profile generator will encode a transform in a BToAxTag that is intentionally different than the inverse of the AToBxTag. Such a profile may still conform to the ICC specification, but to avoid unexpected results it is recommended that such non-invertible transforms are only generated where required for a particular application and that the user is made aware of the non-invertibility of the profile, possibly through the profile description.

32.2 Using a Roundtrip Test to Evaluate Profile Quality

By performing a roundtrip test on a set of sample colors within the effective gamut, it is possible to obtain an indication of the accuracy of inversion in a profile. This test can be carried out for all rendering intents.

If an AToB1Tag is based on measurement data for values in the device encoding and represents an accurate conversion from data encoding to PCS (as required by the specification), then a roundtrip test can also be used to deduce the accuracy of the BToA1Tag.

Independently of the transform accuracy, the accuracy of inversion may also indicate the relative smoothness of the transform.

Figure 32.1 Roundtrip test for colorimetric intent

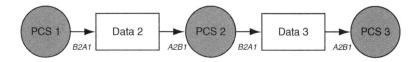

Figure 32.2 Roundtrip test for perceptual and saturation intents in v4 profiles

A valid roundtrip test would be to start with a test set of colors in the data encoding and convert them between data encoding and PCS as shown in Figure 32.1.

The evaluation consists of comparing the PCS 2 values with PCS 3, the additional steps in computing PCS 1, and CMYK 2 values being required to ensure that only colors inside the effective gamut of the data encoding remain in the test set in PCS 2. Ideally the out-of-gamut colors should be removed from the test set to avoid weighting the test with a large proportion of test colors on or close to the gamut boundary.

For v4 profiles containing tags indicating the use of the Perceptual Reference Medum Gamut for either the perceptual or saturation intents, the effective gamut is the PRMG. A valid roundtrip test for the intents where the PRMG is indicated would be to start with colors sampling the PRMG and roundtrip them as shown in Figure 32.2. If not using the PRMG, the actual effective gamut should be used for the roundtrip test, as shown in Figure 32.1.

More details of evaluating v4 profiles is given in Chapter 33. A MATLAB function to perform a roundtrip test can be downloaded from the resources area on the ICC web site.

33

Evaluating Color Transforms in ICC Profiles

ICC input and output profiles contain transforms between device data encodings and the ICC PCS. These transforms should be either accurate or pleasing, depending on the chosen rendering intent.

The following recommendations are provided to assist in the evaluation of the colorimetric and perceptual rendering intent transforms in ICC v4 profiles. Any tolerances provided are guidelines and may not be suitable for all applications.

Three types of test can be considered:

1. Roundtrip tests determine the accuracy with which a given rendering intent within a profile can be inverted, such that when a PCS value is converted to the device encoding and back to the PCS, the difference is minimized. Roundtrip tests are applicable to all intents.
2. Device model tests can determine the accuracy with which the profile predicts the colorimetry of a given device encoding value, or the accuracy with which the profile predicts the device encoding value required to produce a given colorimetry. Device model tests are generally applicable to colorimetric intents.
3. Subjective tests evaluate how pleasing a rendering or re-rendering transform is. Such tests are generally applicable to perceptual intents.

A procedure for applying these tests is listed below. It should be noted that in general preview tags should not be used for testing roundtripping errors. Roundtripping results may be affected by the device characteristics, the profile, and the CMM. Profiles should be evaluated using several CMMs, preferably those that will be used in practice.

1. Determine if the profile AToB1 and BToA1 transforms and media white point tag accurately reflect the device characteristic (after chromatic adaptation from the actual adopted white to D50, if necessary). This is required for all v4 profiles. The AToB1 transform should be checked by comparing PCS values to device measurements obtained using the actual illumination, chromatically adapted to D50 using the "chad" tag. The BToA1 transform is

Color Management: Understanding and Using ICC Profiles   Edited by Phil Green
© 2010 John Wiley & Sons, Ltd

checked by transforming device values to the PCS (using AToB1), back to device (using BToA1), and back to PCS (using AToB1), and comparing the first and second PCS values.

2. Determine if the profile media-relative colorimetric and perceptual transforms are identical. If they are identical the profile contains no color rendering or re-rendering – proceed to step 5. If they are not identical, proceed to step 3.

3. Determine whether the transforms contain three-dimensional CLUTs larger than $2 \times 2 \times 2$ ($2 \times 2 \times 2$ CLUTs are used in some transforms like a matrix).

4a. If the transforms do not contain CLUTs larger than $2 \times 2 \times 2$, they should be tested by roundtripping test colors spanning the profile gamut, or the Perceptual Reference Medium Gamut (PRMG) if its use is indicated. The roundtripping should be very accurate, since the transforms are analytical and there is no need for color rendering or re-rendering trade-offs. Roundtripping color differences (in CIELAB ΔE_{ab}^*) should be less than 0.5 mean and less than 1 max.

It should be noted that the gamut for a perceptual transform can be larger than the PRMG even when use of the PRMG is indicated. An image can be color rendered or re-rendered appropriately for the PRMG, but still be encoded in a color space which has a larger gamut than the PRMG.

4b. If the transforms do contain CLUTs larger than $2 \times 2 \times 2$, they should be tested by roundtripping colors as described in 4a, but the accuracy requirements will by necessity be less stringent, because of interpolation of the CLUTs. Roundtripping color differences in CIELAB ΔE_{ab}^* should be less than 1 mean and less than 3 maximum.

5a. If the media-relative colorimetric and perceptual transforms are not identical and the use of the PRMG is indicated, some additional leeway in the roundtripping accuracy should be granted to allow for reasonable trade-offs between color rendering/re-rendering and spanning the PRMG. Transforms should be tested by roundtripping colors spanning the PRMG, but the roundtripping color differences within the PRMG should be less than 2 mean and less than 10 maximum.

5b. If the media-relative colorimetric and perceptual transforms are not identical and the use of the PRMG is not indicated, knowledge of the target gamut for the perceptual color rendering or re-rendering is necessary to apply the objective evaluation methods outlined above to the perceptual rendering intent transforms. However, subjective evaluation methods can be used even when such knowledge is not available.

Version 4 profiles with no color rendering or re-rendering indicate that the images to which they are assigned are to be interpreted as being already color rendered appropriately for the PRM. Such profiles can be evaluated objectively, so long as any errors are very small. However, visual evaluation is useful to determine the degree to which in-gamut colors match when produced using different printers. Very small CIELAB color differences can sometimes result in a visual mismatch, particularly for near-neutral colors. Visual evaluation can also help identify measurement or illumination errors.

Version 4 profiles that contain color rendering or re-rendering intentionally modify the output colorimetry and hence cannot be completely evaluated using only objective methods. Subjective evaluation using large numbers of real images is required to determine the quality of the color rendering or re-rendering.

To subjectively evaluate the AToB0 transform in a profile for use as a source profile, a collection of images that are judged to be of excellent quality when interpreted directly according to the source color encoding should be printed on a print medium with a color gamut similar to the PRMG using the perceptual rendering intent, and then viewed in the PRM viewing conditions.

To subjectively evaluate the BToA0 transform in a profile for use as a destination profile, a collection of images that are judged to be of excellent quality when printed and viewed using a print medium and viewing conditions similar to those of the PRM should be reproduced on the destination medium using the perceptual rendering intent, and then viewed in the intended destination viewing conditions.

Images suitable for subjective BToA0 transform evaluation are provided in ISO 12640-3, and can also be made from camera raw files by carefully color rendering into ROMM RGB and evaluating the results by printing colorimetrically on a print medium with a color gamut similar to the PRMG, and then viewing the prints in the PRM viewing conditions.

For improved interoperability with v2 source profiles, it may also be desirable to subjectively evaluate BToA0 transforms using images that are in color encodings with reference media different from the PRM (and have v2 profiles embedded). Such co-optimization is desirable when it can be achieved without sacrificing the subjective quality of the PRM image reproduction.

34

Profile Compliance Testing with SampleICC

34.1 Introduction

The ICC profile format specification defines an open file format that acts as an exchangeable container for data that is used to perform color transformations. The file format defines data elements, order, and meaning, but the content of these elements will depend on the goals of the profile creator. Determining correctness of a profile's generation or modification as outlined by the profile specification can go beyond the information provided directly by the profile specification. This is because the exact tests to perform, the order in which they should be performed, and the interpretation of their results are not clearly defined by the profile specification.

Because the tests and interpretation of the results can be context specific, providing a testing specification is not straightforward. Furthermore, any changes to the profile specification would require changes to the testing specification, and keeping them synchronized becomes a problem.

Rather than attempt to provide a complete profile testing specification, this chapter will describe some of the issues of profile conformance testing that were identified and addressed as part of implementing profile conformance capabilities into the SampleICC project.

The SampleICC project (see http://sampleicc.sourceforge.net) is an open source object-oriented C++ development effort that was written to provide an example of how various aspects of color management can be implemented. It is maintained by the Architecture Working Group of the ICC.

This chapter provides overviews of test types and specifics of some important tests that are implemented in SampleICC's IccProfLib. It is hoped that this discussion, together with the code in SampleICC, can be used as a guide to understanding issues related to determining profile compliance. However, neither this chapter nor the code in SampleICC should be considered as a profile conformance testing specification.

Color Management: Understanding and Using ICC Profiles   Edited by Phil Green
© 2010 John Wiley & Sons, Ltd

Not all tests related to profile compliance are presented in this chapter. Such tests can be determined through a careful study of the ICC profile specification. Even though the tests in SampleICC are extensive, neither the SampleICC code nor the chapter can substitute for the ICC profile specification, which should always be considered the ultimate source in determining profile compliance.

34.2 Profile Compliance Levels

In general there are three questions that can be asked in relation to profile compliance:

1. Is the profile legible? (Can the profile be read?)
2. Does the profile conform to the specification?
3. Is the profile usable?

The first question relates to whether the ICC profile can be parsed.

The second question assumes the first to be true. Once one can parse the file then there is a question whether the data within the file makes logical sense (as defined by the profile specification).

The third question is the most important one, and often goes beyond simple profile conformance testing. In certain cases the answer to the second question may be negative and the answer to the third affirmative. A profile might not comply with the specification in some area outside the scope where the profile will be used, and thus the profile may be usable.

The issue of profile usability can partially be addressed by introducing levels of compliance. This document considers four levels of compliance:

1. Compliant – The profile is completely compliant.
2. Warning – Possible problems exists, but the profile is compliant with the specification.
3. Non-compliant – The profile does not strictly follow the ICC profile format specification.
4. Critical error – The non-compliance of the profile makes the profile unusable.

Notice there is a distinction between levels 3 and 4. Both levels indicate that the profile does not strictly follow the ICC profile format specification. For example, there are non-compliant issues that may not effect whether a CMM (or profile user) can successfully understand and apply or use the profile under specific conditions. The distinction between levels 3 and 4 in many cases can be argued one way or another. For example, a non-compliant issue for a CMM might be a critical error to a profile manager and vice versa.

It should be noted that the use of compliance levels in this chapter applies to the successful application of a profile within SampleICC.

The distinction of compliance levels also allows for a discussion of robustness to allow for future extensions of the profile specification. For example, at some future time an informational-only field could conceivably be added using part of the reserved portion of the profile header. This would make the profile non-compliant in terms of an older specification, but since the field is informational it would not be a critical error from a profile application point of view.

There is a clear distinction between profile compliance and validity. This chapter only considers issues of compliance. Just as a profile can be non-compliant but usable, it is possible for a profile to be completely compliant to the ICC profile specification and yet not be valid for any constructive use. For example, a profile that turns all images black could be labeled as compliant if it passes all the types of tests involved in this document.

34.3 Profile Compliance Testing in SampleICC

The profile object hierarchy of a profile loaded into memory is represented in Figure 34.1.

The **CIccProfile** class is responsible for defining operations on profiles. A **CIccProfile** object essentially maintains two data members – the profile header data and a list of **IccTagEntries**.

Each **IccTagEntry** contains a pointer to a **CIccTag**-derived object as well as a **TagInfo** structure that contains the tag signature, size, and location of the tag in a loaded file.

All tag types defined by the ICC specification are represented by **CIccTag**-derived classes.

The SampleICC project splits answering the profile compliance questions into two parts. These two parts are implemented through the use of two public member functions of the **CIccProfile** class:

1. **CIccProfile::ReadValidate().** The **CIccProfile::ReadValidate()** member function directly answers the first of the profile compliance questions. It is an alternative version of the **CIccProfile::Read()** function which parses an ICC profile file and generates the object structure in memory. While parsing the ICC profile it contributes to a profile compliance report and returns the maximum compliance level of all the tests that are performed.

   The following operations are performed in **ReadValidate()**. Each of these operations involves making calls to protected member functions of **CIccProfile**. The tests/

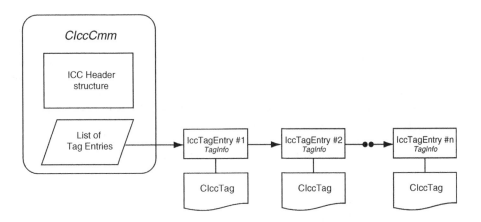

Figure 34.1   Object hierarchy of a profile in memory

operations are:

(a) The header is read and the tag directory is parsed and read.

(b) The actual file size is compared to the profile size specified in the header.

(c) The profile ID is calculated and compared to that found in the header.

(d) Each tag defined by the profile tag directory is parsed and read in using the tag classes in SampleICC's IccProfLib.

2. **CIccProfile::Validate().** The **CIccProfile::Validate()** member function tries to answer the last two profile compliance questions. Issues of usability (expressed through compliance-level reporting) are limited to the known ability of the **CIccCmm** class to apply the profile. Whether a profile is useable for any other specific purpose goes beyond the scope of this function.

The **CIccProfile::Validate()** member function acts separately from the **CIccProfile::ReadValidate()** member function and can be called to determine profile conformance issues of profiles that are being generated in memory using SampleICC's IccProfLib before they are written to file.

The **CIccProfile::Validate()** function also contributes to a profile compliance report and returns the maximum compliance level of all the tests that are performed.

The following operations are performed in **CIccProfile::Validate()**. Each of these operations involves making calls to protected member functions of CIccProfile. The tests/operations are:

1. The header is checked for compliance with the profile specification. (See **CIccProfile::CheckHeader()** in SampleICC's IccProfLib.)

2. Tags are checked for uniqueness. (See **CIccProfile::AreTagsUnique()** in SampleICC's IccProfLib.)

3. Required tags are checked for the profile type (based upon profile type information stored in the header). (See **CIccProfile::CheckRequiredTags()** in SampleICC's IccProfLib.)

4. The tag types are checked. (See **CIccProfile::CheckTagTypes()** in SampleICC's IccProfLib.)

5. The virtual **CIccTag::Validate()** member function for each of the **CIccTag**-derived objects associated with the profile's Tag directory is called. Within IccProfLib, each **CIccTag** class is responsible for providing its own specification compliance testing.

Note that "private" tags created by additional tag factories (based on the **CIccTagFactory** interface) registered through the singleton CIccTagCreator tag factory system can still provide their own internal validation by implementing the tag's virtual Validate() function. However, the tests in **CIccProfile::CheckRequiredTags()** and **CIccProfile::CheckTagTypes()** are unaware of any needs or requirements of such "private" tags.

SampleICC's IccProfLib also provides a single static global function (**ValidateIccProfile**) to simplify profile compliance testing. This function performs the following operations:

1. It initiates profile file input/output.

2. It allocates a new **CIccProfile** object.

3. It calls **CIccProfile::ReadValidate()**.

4. It calls **CIccProfile::Validate()**.
5. Steps 3 and 4 generate a profile file report and compliance level for the profile.
6. It returns the loaded profile object structure.

SampleICC's profile compliance testing is incorporated into the ProfileDump utility, and a compiled version of this utility is available for the convenience of users in the profile viewing and testing section of the ICC web site.

Index

Printed and bound by CPI Group (UK) Ltd, Croydon, CR0 4YY